SO-AQL-732

STUDENT LECTURE NOTEBOOK

Tarbuck • Lutgens

E A R T H

An Introduction to Physical Geology

EIGHTH EDITION

PEARSON

Prentice
Hall

Upper Saddle River, NJ 07458

Editor-in-Chief, Science: John Challice
Executive Editor: Patrick Lynch
Assistant Editor: Melanie Cutler
Vice President of Production & Manufacturing: David W. Riccardi
Executive Managing Editor: Kathleen Schiaparelli
Assistant Managing Editor: Becca Richter
Production Editor: Elizabeth Klug
Supplement Cover Management/Design: Paul Gourhan
Manufacturing Buyer: Ilene Kahn
Cover Photo Credit: Hiking over a sandstone wave, Paria Canyon, Vermilion Cliffs Wilderness, Arizona (Galen Rowell/Mountain Light Photography, Inc.)

© 2005 Pearson Education, Inc.
Pearson Prentice Hall
Pearson Education, Inc.
Upper Saddle River, NJ 07458

All rights reserved. No part of this book may be reproduced in any form or by any means, without permission in writing from the publisher.

Pearson Prentice Hall® is a trademark of Pearson Education, Inc.

The author and publisher of this book have used their best efforts in preparing this book. These efforts include the development, research, and testing of the theories and programs to determine their effectiveness. The author and publisher make no warranty of any kind, expressed or implied, with regard to these programs or the documentation contained in this book. The author and publisher shall not be liable in any event for incidental or consequential damages in connection with, or arising out of, the furnishing, performance, or use of these programs.

Printed in the United States of America

10 9 8 7 6 5 4 3 2

ISBN 0-13-144729-7

Pearson Education Ltd., *London*
Pearson Education Australia Pty. Ltd., *Sydney*
Pearson Education Singapore, Pte. Ltd.
Pearson Education North Asia Ltd., *Hong Kong*
Pearson Education Canada, Inc., *Toronto*
Pearson Educación de Mexico, S.A. de C.V.
Pearson Education—Japan, *Tokyo*
Pearson Education Malaysia, Pte. Ltd.

Contents

To the Student

This *Student Lecture Notebook* is designed to help you do your best in this physical geology course.

Key images from the textbook and every illustration from the Instructor's Transparency Set are reproduced in this notebook. Because you won't have to redraw the art in class, you can focus your attention on the lecture, annotate the art, and take your notes in this book.

Leave all your notes together or remove them for integration into a binder with other course materials.

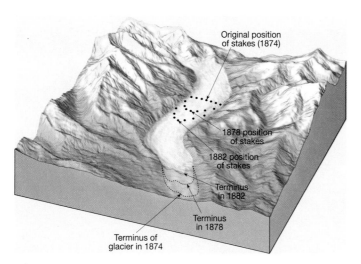

Figure 1.B Ice movement and Rhône Glacier.

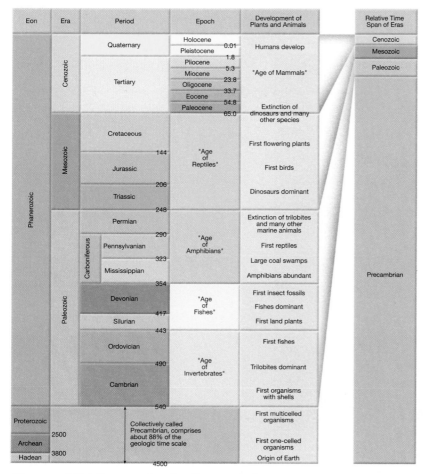

Figure 1.7 The geologic time scale.

Tarbuck/Lutgens, *Earth: An Introduction to Physical Geology, 8e*

© 2005 Pearson Prentice Hall, Inc.

NOTES:

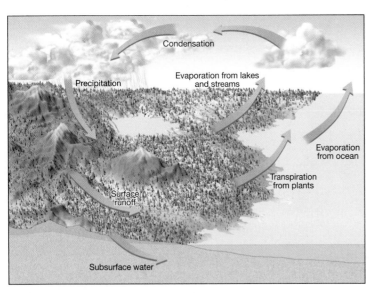

Figure 1.11 The hydrologic cycle.

Figure 1.13 A,B,C,D,E Formation of the solar system: nebular hypothesis.

Tarbuck/Lutgens, *Earth: An Introduction to Physical Geology, 8e*
© 2005 Pearson Prentice Hall, Inc.

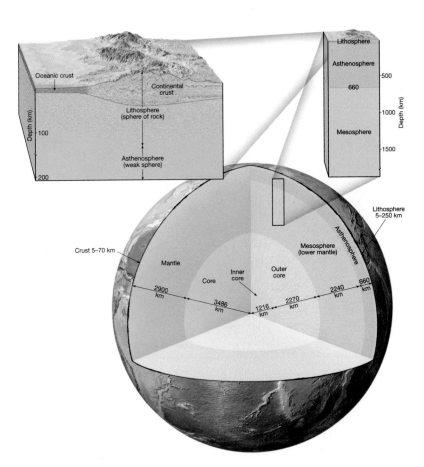

Figure 1.14 Earth's layered structure.

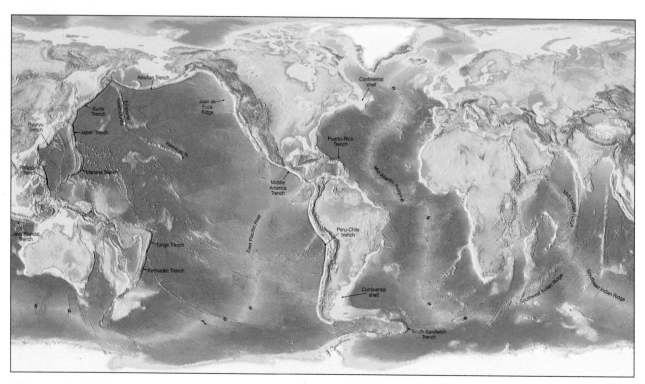

Figure 1.15 The topography of Earth's solid surface.

NOTES:

© 2005 Pearson Prentice Hall, Inc.

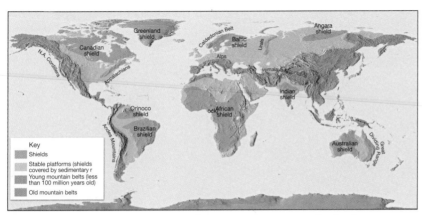

Figure 1.16 Distribution of Earth's mountain belts, stable platforms, and shields.

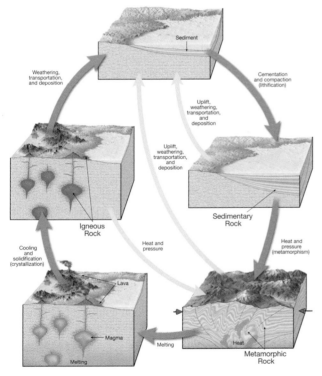

Figure 1.21 The rock cycle.

© 2005 Pearson Prentice Hall, Inc.

NOTES:

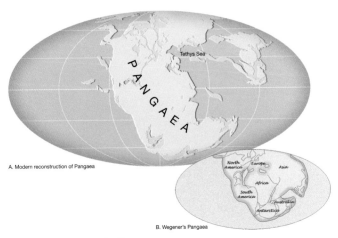

Figure 2.2 A,B Pangaea.

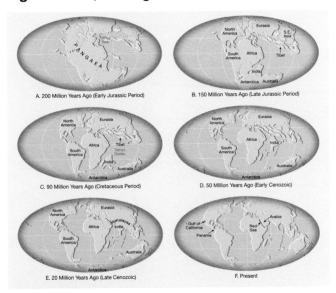

Figure 2.A Several views of the breakup of Pangaea.

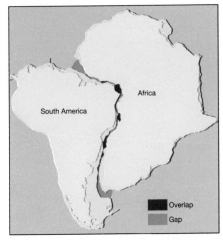

Figure 2.3 Best fit of South America and Africa.

Tarbuck/Lutgens, *Earth: An Introduction to Physical Geology, 8e*

© 2005 Pearson Prentice Hall, Inc.

NOTES:

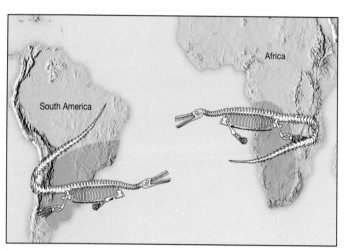

Figure 2.4 Fossils of *Mesosaurus.*

Figure 2.6 Matching mountain ranges across the North Atlantic.

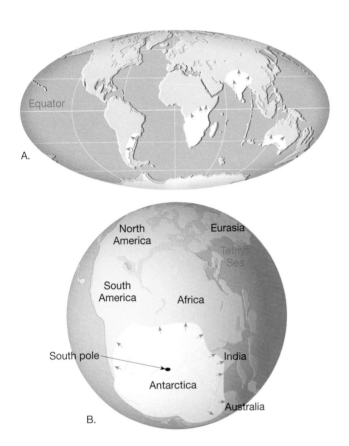

Figure 2.7 A,B Paleoclimatic evidence for continental drift.

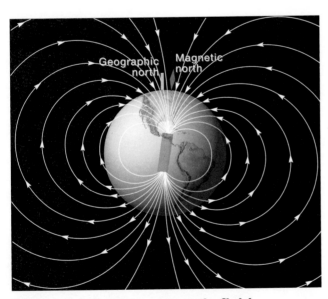

Figure 2.8 Earth's magentic field.

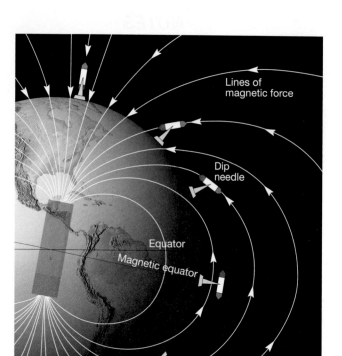

Figure 2.9 Earth's magnetic field causes a dip needle to align with lines of magnetic force.

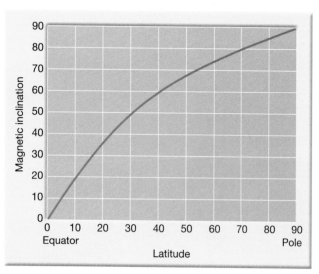

Figure 2.10 Magnetic inclination and corresponding latitude.

© 2005 Pearson Prentice Hall, Inc.

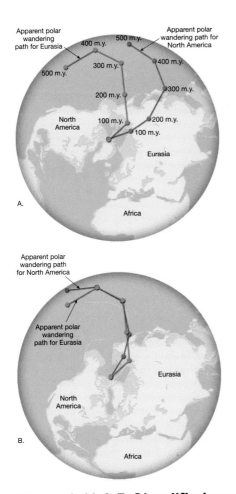

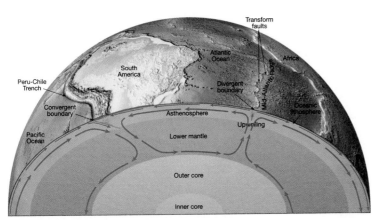

Figure 2.11 A,B Simplified apparent polar-wandering paths.

Figure 2.12 Seafloor spreading.

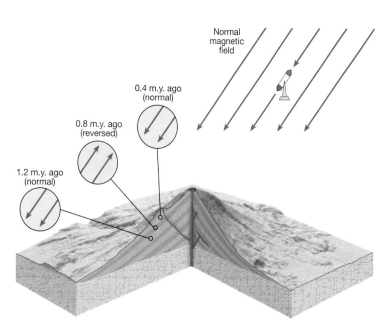

Figure 2.13 Paleomagnetism preserved in lava flows of different ages.

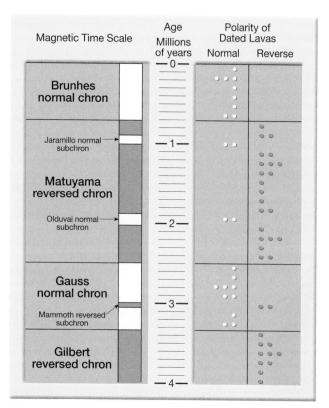

Figure 2.14 Time scale of Earth's magnetic field.

© 2005 Pearson Prentice Hall, Inc.

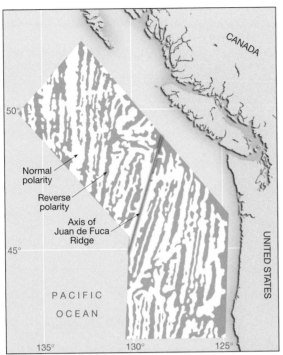

Figure 2.15 Pattern of alternating stripes of high- and low-intensity magnetism.

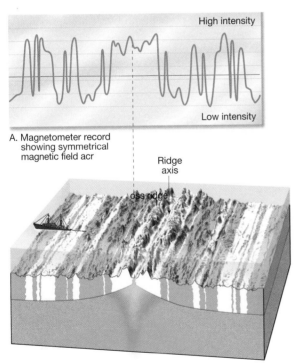

A. Magnetometer record showing symmetrical magnetic field acr

Ridge axis

oss ridge

B. Research vessel towing magnetometer across ridge crest

Figure 2.16 A,B The ocean floor as a magnetic tape recorder.

Tarbuck/Lutgens, *Earth: An Introduction to Physical Geology, 8e*
© 2005 Pearson Prentice Hall, Inc.

NOTES:

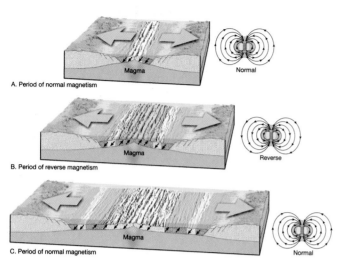

Figure 2.17 A,B,C *New basalt is magnetized according to Earth's existing magnetic field.*

Figure 2.18 *An alternate hypothesis to continental drift: an expanding Earth.*

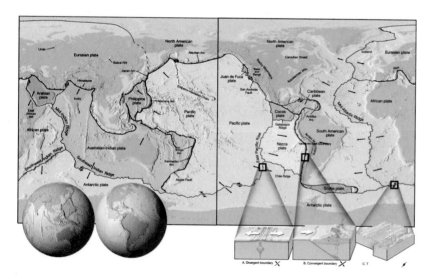

Figure 2.19 *A mosaic of rigid plates constitutes Earth's outer shell.*

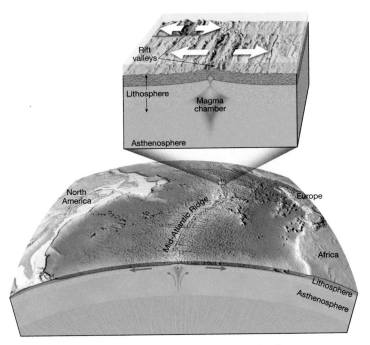

Figure 2.20 Divergent plate boundaries.

NOTES:

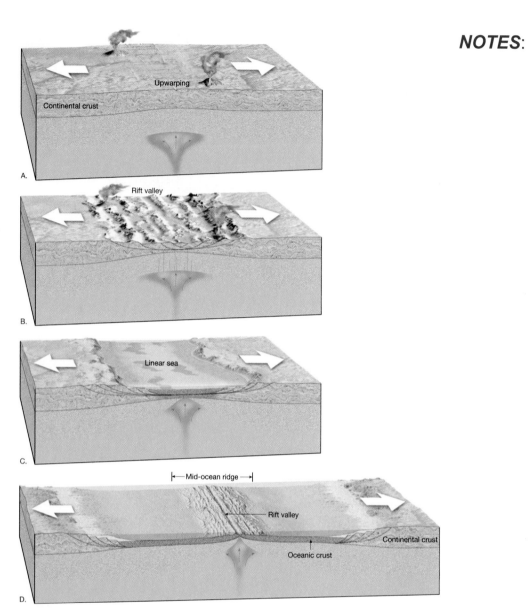

Figure 2.21 A,B,C,D Continental rifting and the formation of a new ocean basin.

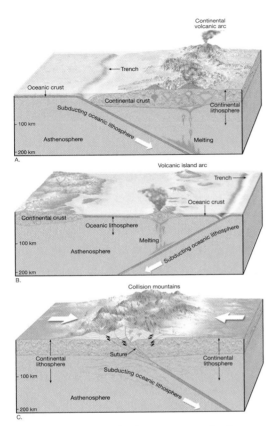

Figure 2.22 A,B,C *Zones of plate convergence.*

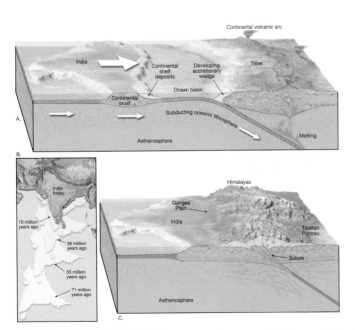

Figure 2.23 A,B,C *The ongoing collision of India and Asia.*

Tarbuck/Lutgens, *Earth: An Introduction to Physical Geology, 8e*
© 2005 Pearson Prentice Hall, Inc.

NOTES:

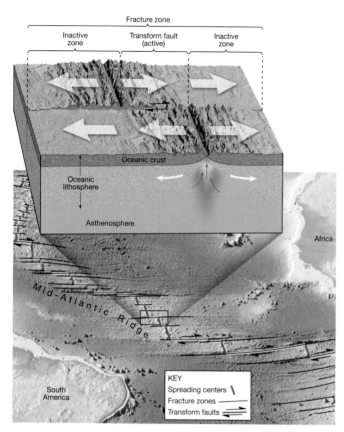

Figure 2.24 The role of transform faults.

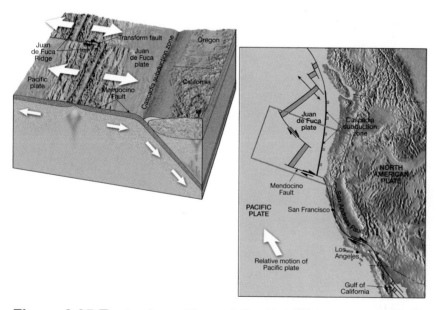

Figure 2.25 Tectonic setting of the Pacific Northwest.

Tarbuck/Lutgens, *Earth: An Introduction to Physical Geology, 8e*
© 2005 Pearson Prentice Hall, Inc.

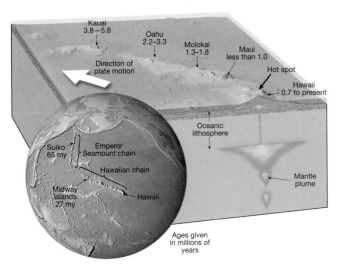

Figure 2.27 *The island chain from Hawaii to the Aleutians is the result of hot spot activity and plate motion.*

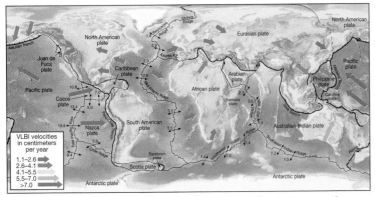

Figure 2.29 *Directions and rates of plate motion.*

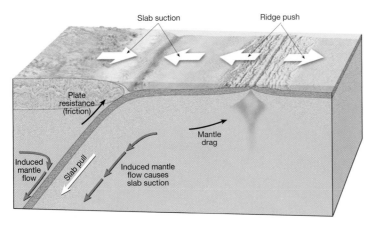

Figure 2.30 *Forces that act on plates.*

© 2005 Pearson Prentice Hall, Inc.

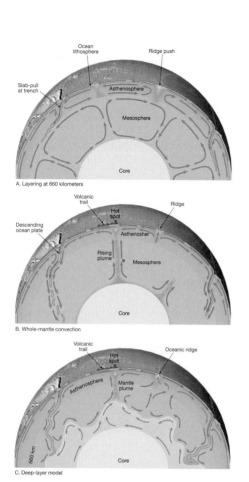

A. Layering at 660 kilometers

B. Whole-mantle convection

C. Deep-layer model

Figure 2.31 A,B,C *Proposed models for mantle convection.*

NOTES:

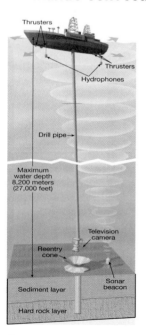

Figure 2.C *Ocean drilling.*

© 2005 Pearson Prentice Hall, Inc.

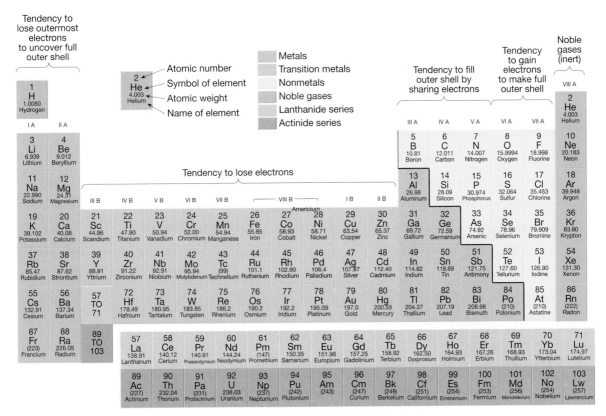

Figure 3.3 *Periodic table of elements.*

NOTES:

NOTES:

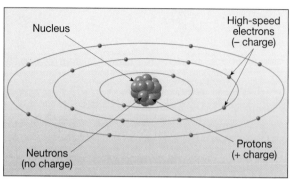

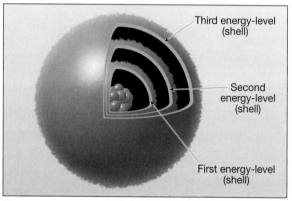

Figure 3.4 A,B Two models of the atom.

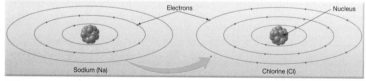

Figure 3.5 Chemical bonding of sodium and chlorine.

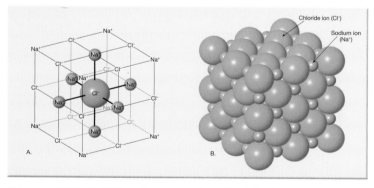

Figure 3.6 A,B Sodium and chloride ions in table salt.

Tarbuck/Lutgens, Earth: An Introduction to Physical Geology, 8e
© 2005 Pearson Prentice Hall, Inc.

NOTES:

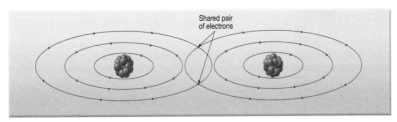

Figure 3.7 Two chlorine atoms share an electron pair.

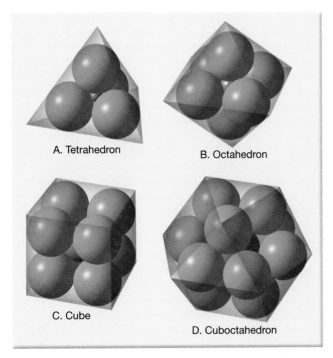

A. Tetrahedron

B. Octahedron

C. Cube

D. Cuboctahedron

Figure 3.8 Ideal geometrical packing for various-sized atoms.

Tarbuck/Lutgens, *Earth: An Introduction to Physical Geology, 8e*
© 2005 Pearson Prentice Hall, Inc.

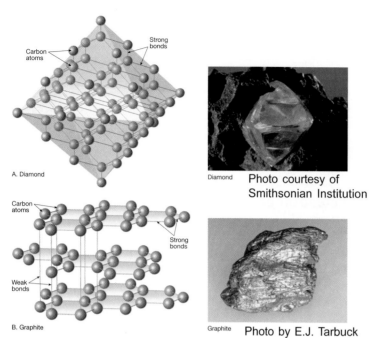

Diamond Photo courtesy of Smithsonian Institution

Graphite Photo by E.J. Tarbuck

Figure 3.10 A. Three-dimensional structure of diamond. B. Graphite: carbon atoms are bonded into sheets.

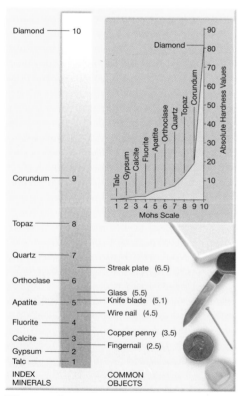

Figure 3.13 Mohs scale of hardness.

© 2005 Pearson Prentice Hall, Inc.

NOTES:

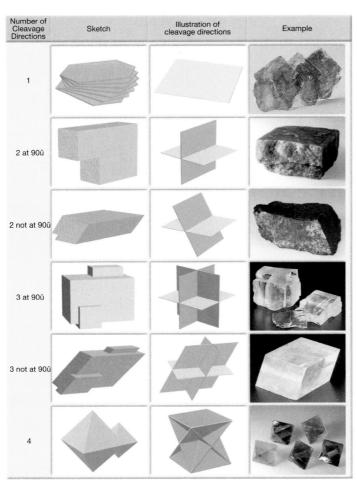

Number of Cleavage Directions	Sketch	Illustration of cleavage directions	Example
1			
2 at 90û			
2 not at 90û			
3 at 90û			
3 not at 90û			
4			

Photos by E.J. Tarbuck

*Figure 3.15 **Common cleavage directions.***

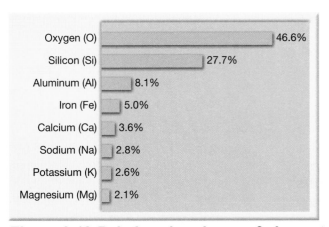

Oxygen (O)	46.6%
Silicon (Si)	27.7%
Aluminum (Al)	8.1%
Iron (Fe)	5.0%
Calcium (Ca)	3.6%
Sodium (Na)	2.8%
Potassium (K)	2.6%
Magnesium (Mg)	2.1%

*Figure 3.18 **Relative abundance of elements in the continental crust.***

© 2005 Pearson Prentice Hall, Inc.

Figure 3.19 A,B Two representations of the silicon-oxygen tetrahedron.

Negative Ion (Anion)	Positive Ions (Cations)		
	Si⁴⁺ 0.39	Al³⁺ 0.51	
	Fe³⁺ 0.64	Mg²⁺ 0.66	Fe²⁺ 0.74
O²⁻ 1.40	Na¹⁺ 0.97	Ca²⁺ 0.99	K¹⁺ 1.33

Figure 3.20 Relative sizes and charges of ions of the most common elements in Earth's crust.

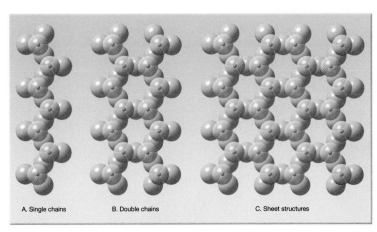

A. Single chains   B. Double chains   C. Sheet structures

*Figure 3.21 A,B,C **Three types of silicate structures.***

Mineral		Idealized Formula	Cleavage	Silicate Structure	
Olivine		(Mg, Fe)$_2$SiO$_4$	None	Single tetrahedron	
Pyroxene group (Augite)		(Mg,Fe)SiO$_3$	Two planes at right angles	Single chains	
Amphibole group (Hornblende)		Ca$_2$ (Fe,Mg)$_5$Si$_8$O$_{22}$(OH)$_2$	Two planes at 60° and 120°	Double chains	
Micas	Biotite	K(Mg,Fe)$_3$AlSi$_3$O$_{10}$(OH)$_2$	One plane	Sheets	
	Muscovite	KAl$_2$(AlSi$_3$O$_{10}$)(OH)$_2$			
Feld-spars	Potassium feldspar (Orthoclase)	KAlSi$_3$O$_8$	Two planes at 90°	Three-dimensional networks	
	Plagioclase	(Ca,Na)AlSi$_3$O$_8$			
Quartz		SiO$_2$	None		

*Figure 3.22 **Common silicate minerals.***

Tarbuck/Lutgens, *Earth: An Introduction to Physical Geology, 8e*
© 2005 Pearson Prentice Hall, Inc.

NOTES:

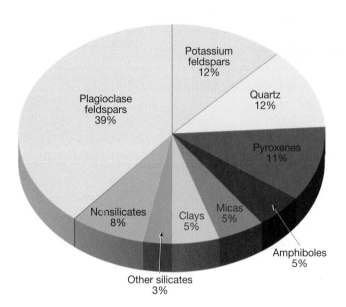

Figure 3.23 Estimated percentages of the most common minerals in Earth's crust.

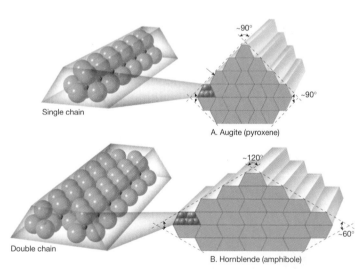

Figure 3.28 Cleavage angles for augite and hornblende.

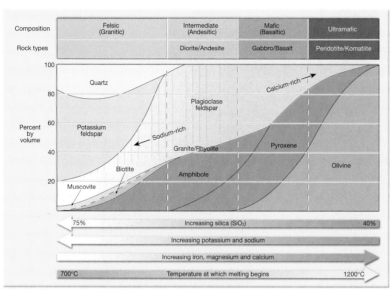

Figure 4.7 Mineralogy of common igneous rocks.

Chemical Composition		Granitic (Felsic)	Andesitic (Intermediate)	Basaltic (Mafic)	Ultramafic
Dominant Minerals		Quartz Potassium feldspar Sodium-rich plagioclase feldspar	Amphibole Sodium- and calcium-rich plagioclase feldspar	Pyroxene Calcium-rich plagioclase feldspar	Olivine Pyroxene
Accessory Minerals		Amphibole Muscovite Biotite	Pyroxene Biotite	Amphibole Olivine	Calcium-rich plagioclase feldspar
TEXTURE	Phaneritic (coarse-grained)	Granite	Diorite	Gabbro	Peridotite
	Aphanitic (fine-grained)	Rhyolite	Andesite	Basalt	Komatiite (rare)
	Porphyritic	"Porphyritic" precedes any of the above names whenever there are appreciable phenocrysts			Uncommon
	Glassy	Obsidian (compact glass) Pumice (frothy glass)			
	Pyroclastic (fragmental)	Tuff (fragments less than 2 mm) Volcanic Breccia (fragments greater than 2 mm)			
Rock Color (based on % of dark minerals)		0% to 25%	25% to 45%	45% to 85%	85% to 100%

Figure 4.8 Classification of igneous rocks.

© 2005 Pearson Prentice Hall, Inc.

NOTES:

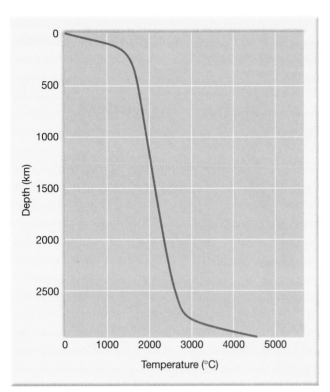

Figure 4.18 Estimated temperature distribution of the crust and mantle.

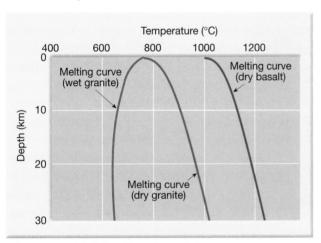

Figure 4.19 Idealized melting temperature curves.

NOTES:

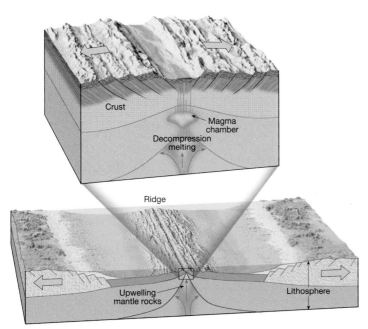

Figure 4.20 Ascension of hot mantle rock.

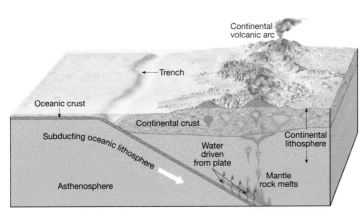

Figure 4.21 An oceanic plate descends into the mantle.

Photo by E.J. Tarbuck

Figure 4.22 Ash and pumice ejected during an eruption of Mt. Mazama.

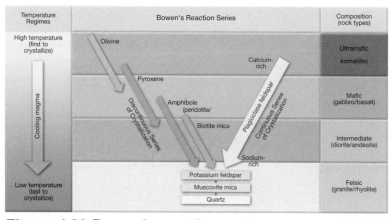

Figure 4.23 Bowen's reaction series.

NOTES:

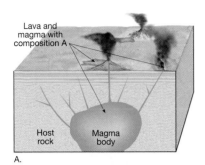

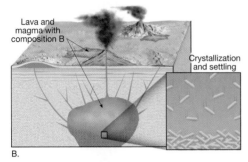

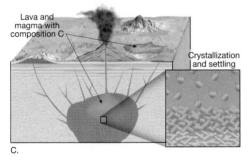

Figure 4.24 A,B,C Illustration of how a magma evolves.

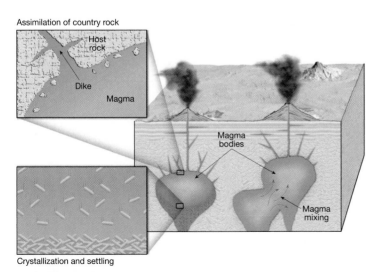

Figure 4.25 Three ways a magma body may be altered.

Tarbuck/Lutgens, *Earth: An Introduction to Physical Geology, 8e*
© 2005 Pearson Prentice Hall, Inc.

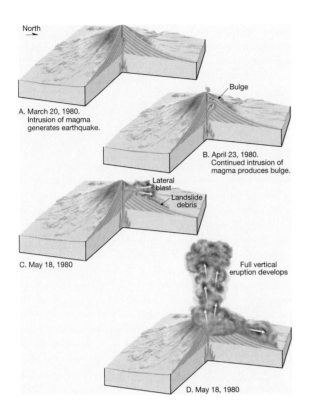

Figure 5.A **The eruption of Mount St. Helens.**

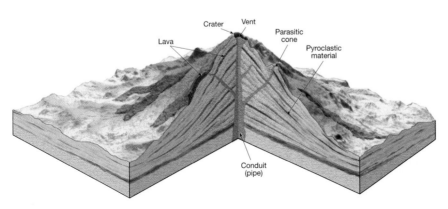

Figure 5.9 **Anatomy of a "typical" composite cone.**

NOTES:

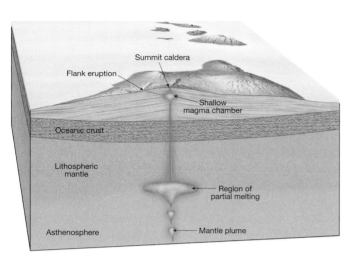

Figure 5.10 A shield volcano.

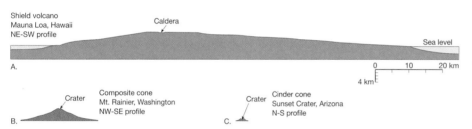

Figure 5.12 A,B,C Profiles of volcanic landforms.

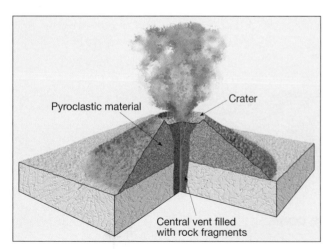

Figure 5.14 Profile of a cinder cone.

Tarbuck/Lutgens, *Earth: An Introduction to Physical Geology, 8e*
© 2005 Pearson Prentice Hall, Inc.

NOTES:

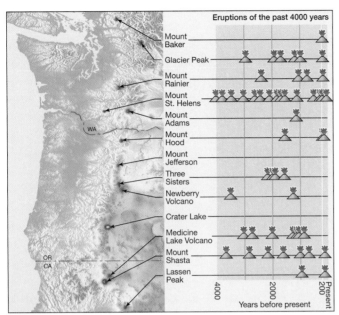

Figure 5.17 Cascade Range volcanoes and eruptions.

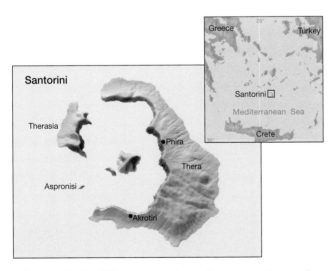

Figure 5.18 Map showing the remains of Santorini.

NOTES:

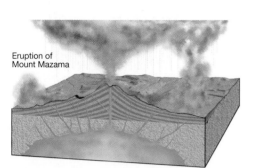

Eruption of
Mount Mazama

Partialy emptied
magma chamber

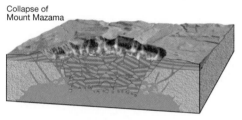

Collapse of
Mount Mazama

Formation of Crater Lake and Wizard Island

*Figure 5.22 Sequence of
events that formed Crater
Lake, Oregon.*

© 2005 Pearson Prentice Hall, Inc.

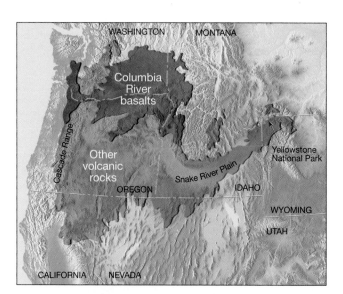

Figure 5.24 *Volcanic areas in the north-western United States.*

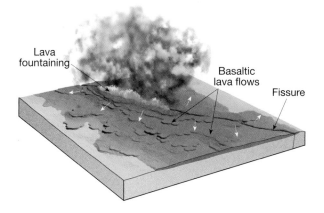

Figure 5.25 A *A basaltic fissure eruption.*

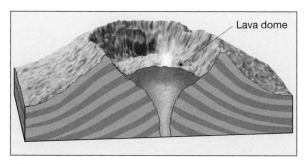

Figure 5.26 *A lava dome develops on Mount St. Helens.*

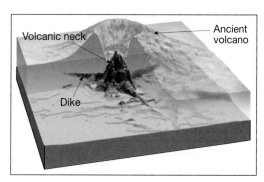

Figure 5.27 Volcanic neck and dike.

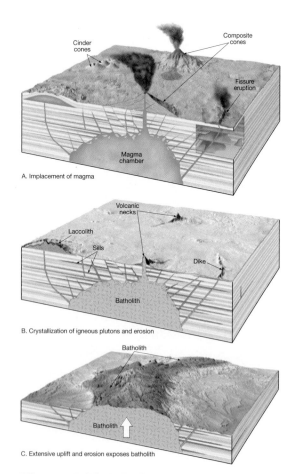

Figure 5.28 A,B,C Basic igneous structures.

NOTES:

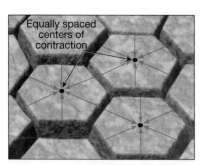

Figure 5.31 *Columnar jointing.*

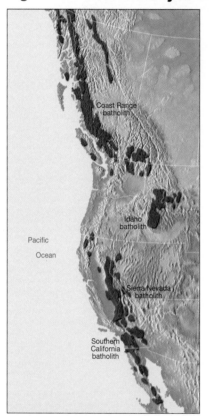

Figure 5.32 *Granitic batholiths in western North America.*

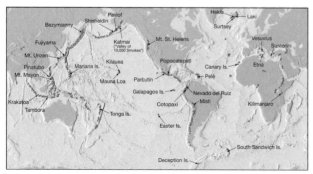

Figure 5.34 *Locations of some of Earth's major volcanoes.*

Tarbuck/Lutgens, *Earth: An Introduction to Physical Geology, 8e*
© 2005 Pearson Prentice Hall, Inc.

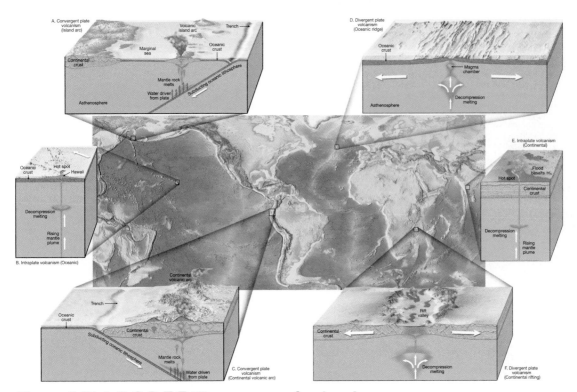

*Figure 5.35 A,B,C,D,E,F **Three zones of volcanism.***

NOTES:

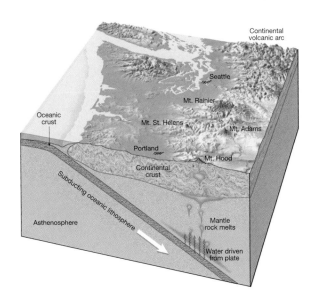

Figure 5.36 Subduction and the Cascade Range.

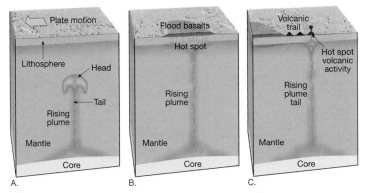

Figure 5.37 A,B,C Model of a mantle plume and hot-spot volcanism.

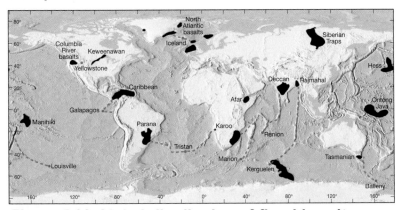

Figure 5.39 Global distribution of flood basalt provinces.

© 2005 Pearson Prentice Hall, Inc.

NOTES:

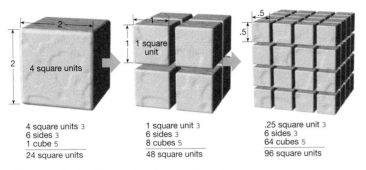

4 square units 3
6 sides 3
1 cube 5
_____
24 square units

1 square unit 3
6 sides 3
8 cubes 5
_____
48 square units

.25 square unit 3
6 sides 3
64 cubes 5
_____
96 square units

Figure 6.2 Chemical weathering.

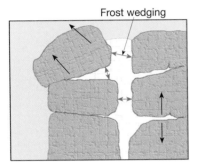

Figure 6.3 Frost wedging.

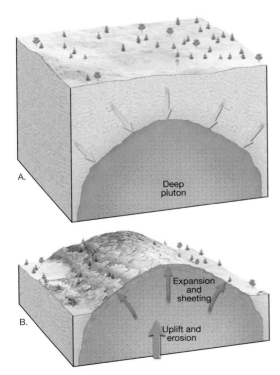

Figure 6.4 A,B Sheeting.

Tarbuck/Lutgens, *Earth: An Introduction to Physical Geology, 8e*
© 2005 Pearson Prentice Hall, Inc.

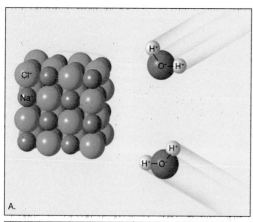

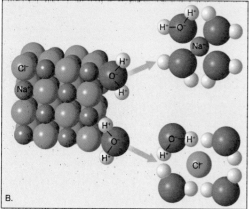

*Figure 6.9 A,B **Halite dissolving in water.***

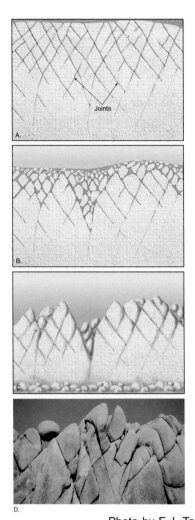

Photo by E.J. Tarbuck

*Figure 6.12 A,B,C,D **Spheroidal weathering of extensively jointed rock.***

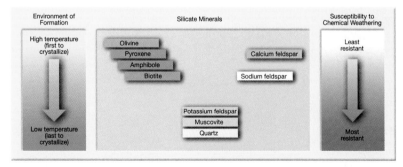

*Figure 6.15 **Weathering of common silicate minerals.***

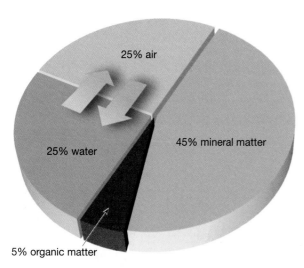

*Figure 6.17 **Composition of a soil in good condition for plant growth.***

*Figure 6.18 **Residual soils and transported soils.***

© 2005 Pearson Prentice Hall, Inc.

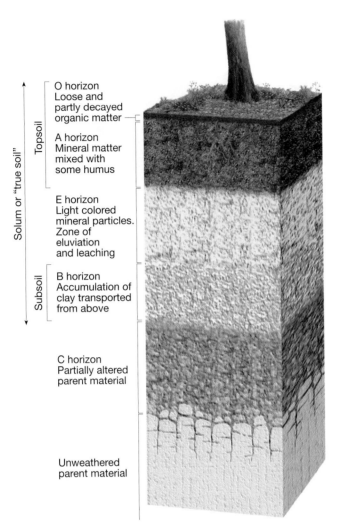

O horizon
Loose and
partly decayed
organic matter

A horizon
Mineral matter
mixed with
some humus

E horizon
Light colored
mineral particles.
Zone of
eluviation
and leaching

B horizon
Accumulation of
clay transported
from above

C horizon
Partially altered
parent material

Unweathered
parent material

Topsoil

Subsoil

Solum or "true soil"

Figure 6.21 Idealized soil profile.

NOTES:

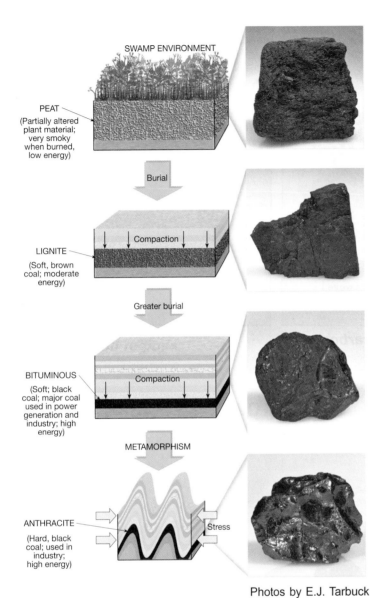

SWAMP ENVIRONMENT

PEAT
(Partially altered
plant material;
very smoky
when burned,
low energy)

Burial

Compaction

LIGNITE
(Soft, brown
coal; moderate
energy)

Greater burial

BITUMINOUS
(Soft; black
coal; major coal
used in power
generation and
industry; high
energy)

Compaction

METAMORPHISM

ANTHRACITE
(Hard, black
coal; used in
industry;
high energy)

Stress

Photos by E.J. Tarbuck

Figure 7.15 Successive stages in the formation of coal.

Detrital Sedimentary Rocks		
Clastic Texture Particle Size	Sediment Name	Rock Name
Coarse (over 2 mm)	Gravel (Rounded particles)	Conglomerate
Coarse (over 2 mm)	Gravel (Angular particles)	Breccia
Medium (1/16 to 2 mm)	Sand (If abundant feldspar is present the rock is called Arkose)	Sandstone
Fine (1/16 to 1/256 mm)	Mud	Siltstone
Very fine (less than 1/256 mm)	Mud	Shale

Chemical Sedimentary Rocks			
Composition	Texture	Rock Name	
Calcite, CaCO$_3$	Nonclastic: Fine to coarse crystalline	Crystalline Limestone	
Calcite, CaCO$_3$	Nonclastic: Fine to coarse crystalline	Travertine	
Calcite, CaCO$_3$	Clastic: Visible shells and shell fragments loosely cemented	Coquina	Biochemical Limestone
Calcite, CaCO$_3$	Clastic: Various size shells and shell fragments cemented with calcite cement	Fossiliferous Limestone	Biochemical Limestone
Calcite, CaCO$_3$	Clastic: Microscopic shells and clay	Chalk	Biochemical Limestone
Quartz, SiO$_2$	Nonclastic: Very fine crystalline	Chert (light colored) Flint (dark colored)	
Gypsum CaSO$_4$•2H$_2$O	Nonclastic: Fine to coarse crystalline	Rock Gypsum	
Halite, NaCl	Nonclastic: Fine to coarse crystalline	Rock Salt	
Altered plant fragments	Nonclastic: Fine-grained organic matter	Bituminous Coal	

Figure 7.17 Identification of sedimentary rocks.

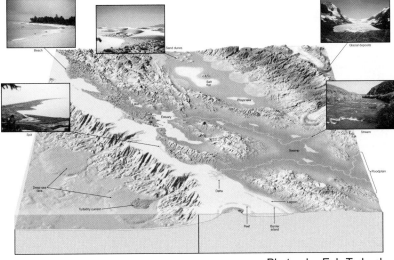

Photos by E.J. Tarbuck

Figure 7.19 Sedimentary environments.

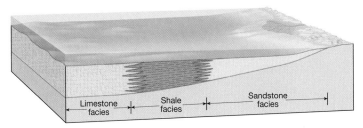

Figure 7.20 Facies.

© 2005 Pearson Prentice Hall, Inc.

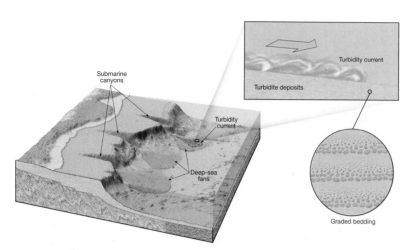

Figure 7.23 *Graded beds.*

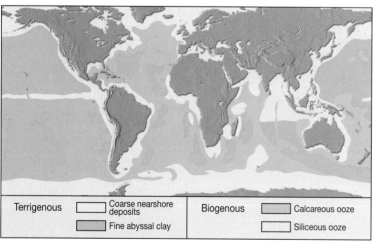

Figure 7.E *Distribution of marine sediment.*

© 2005 Pearson Prentice Hall, Inc.

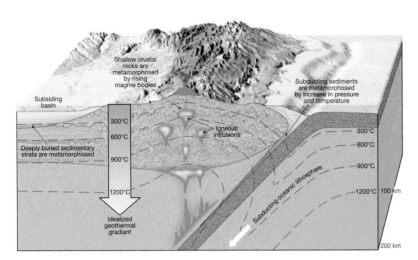

Figure 8.3 *The geothermal gradient and its role in metamorphism.*

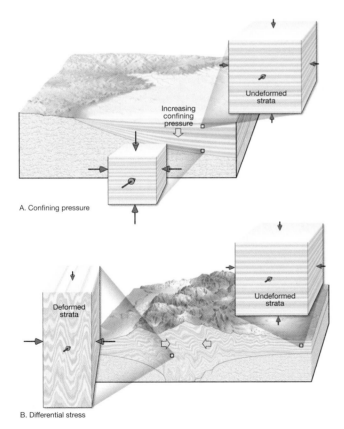

Figure 8.4 A,B *Pressure (stress) as a metamorphic agent.*

NOTES:

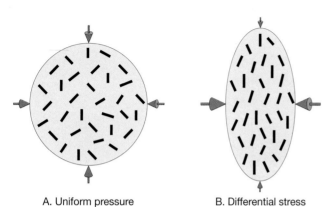

A. Uniform pressure   B. Differential stress

Figure 8.6 A,B Mechanical rotation of platy or elongated mineral grains.

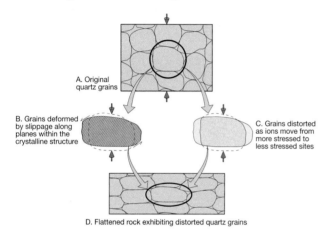

A. Original quartz grains

B. Grains deformed by slippage along planes within the crystalline structure

C. Grains distorted as ions move from more stressed to less stressed sites

D. Flattened rock exhibiting distorted quartz grains

Figure 8.7 Development of preferred orientations of minerals.

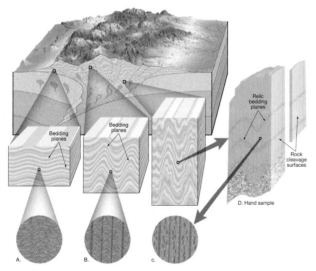

Relic bedding planes

Bedding planes

Bedding planes

Rock cleavage surfaces

D. Hand sample

A.   B.   C.

Figure 8.8 A,B,C,D Development of one type of rock cleavage.

Tarbuck/Lutgens, *Earth: An Introduction to Physical Geology, 8e*
© 2005 Pearson Prentice Hall, Inc.

NOTES:

Rock Name		Texture	Grain Size	Comments	Parent Rock
Slate	Increasing Metamorphism	Foliated	Very fine	Excellent rock cleavage, smooth dull surfaces	Shale, mudstone, or siltstone
Phyllite		Foliated	Fine	Breaks along wavey surfaces, glossy sheen	Slate
Schist		Foliated	Medium to Coarse	Micaceous minerals dominate, scaly foliation	Phyllite
Gneiss		Foliated	Medium to Coarse	Compositional banding due to segregation of minerals	Schist, granite, or volcanic rocks
Migmatite		Foliated	Medium to Coarse	Banded rock with zones of light-colored crystalline minerals	Gneiss, schist
Mylonite		Weakly Foliated	Fine	When very fine-grained, resembles chert, often breaks into slabs	Any rock type
Metaconglomerate		Weakly Foliated	Coarse-grained	Stretched pebbles with preferred orientation	Quartz-rich conglomerate
Marble		Nonfoliated	Medium to coarse	Interlocking calcite or dolomite grains	Limestone, dolostone
Quartzite		Nonfoliated	Medium to coarse	Fused quartz grains, massive, very hard	Quartz sandstone
Hornfels		Nonfoliated	Fine	Usually, dark massive rock with dull luster	Any rock type
Anthracite		Nonfoliated	Fine	Shiny black rock that may exhibit conchoidal fracture	Bituminous coal
Fault breccia		Nonfoliated	Medium to very coarse	Broken fragments in a haphazard arrangement	Any rock type

Figure 8.12 ***Classification of common metamorphic rocks.***

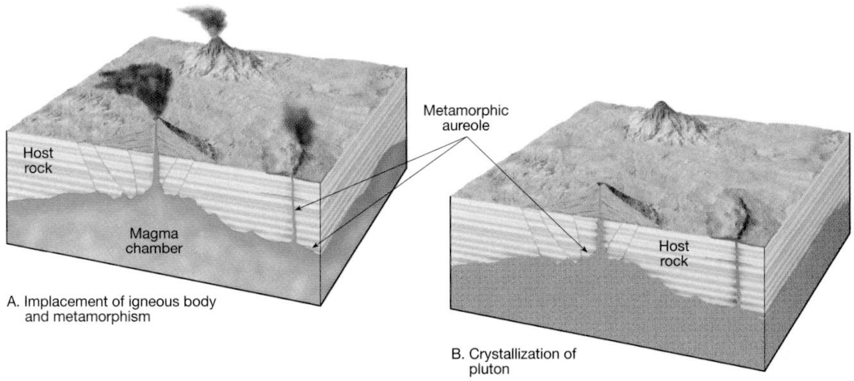

Figure 8.19 A,B ***Contact metamorphism.***

© 2005 Pearson Prentice Hall, Inc.

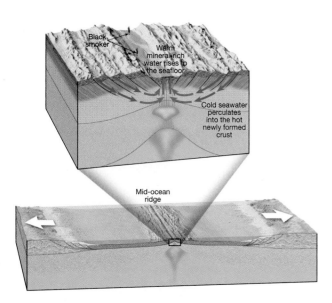

Figure 8.20 *Hydrothermal metamorphism along a mid-ocean ridge.*

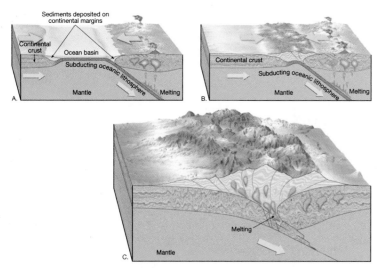

Figure 8.21 A,B,C *Regional metamorphism.*

NOTES:

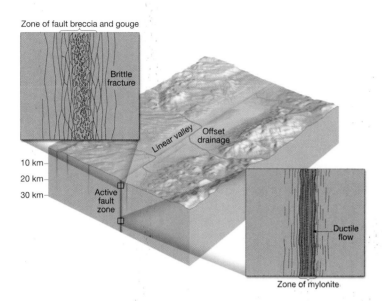

Figure 8.22 *Metamorphism along a fault zone.*

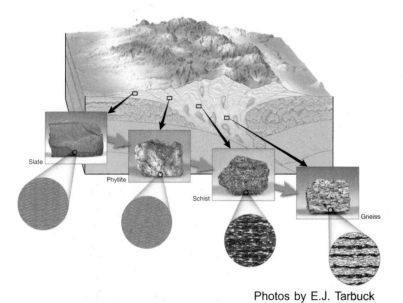

Photos by E.J. Tarbuck

Figure 8.24 *Progressive regional metamorphism.*

© 2005 Pearson Prentice Hall, Inc.

NOTES:

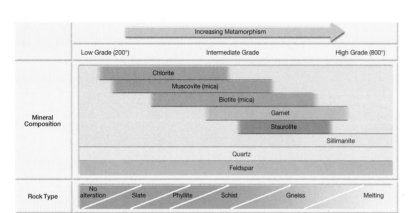

Figure 8.25 Progressive metamorphism of shale.

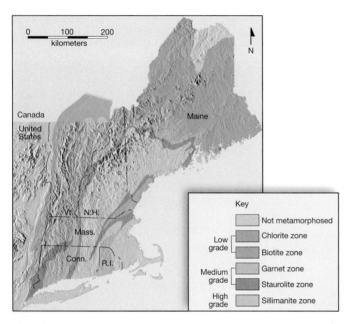

Figure 8.26 Zones of metamorphic intensities in New England.

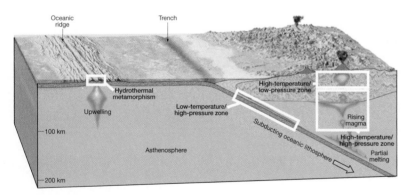

Figure 8.28 Metamorphic environments.

Tarbuck/Lutgens, *Earth: An Introduction to Physical Geology, 8e*
© 2005 Pearson Prentice Hall, Inc.

A.   Photo by E.J. Tarbuck   B.

Figure 9.2 A,B Grand Canyon and the law of superposition.

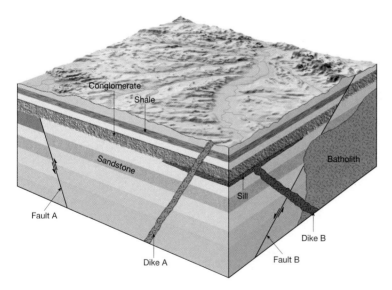

Figure 9.4 Cross-cutting relationships.

NOTES:

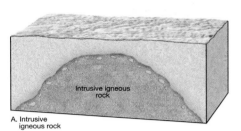

A. Intrusive
igneous rock

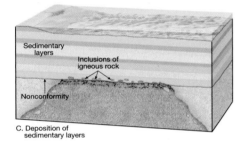

B. Exposure and
weathering of intrusive igneous rock

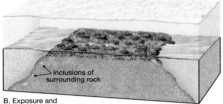

C. Deposition of
sedimentary layers

Figure 9.5 A,B,C Formation of inclusions.

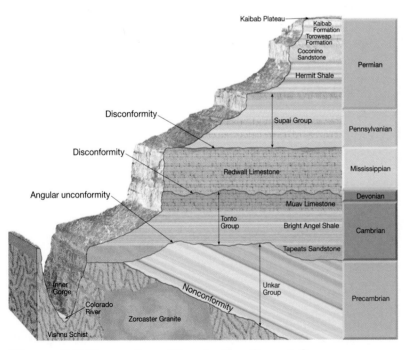

Figure 9.6 Unconformities in the Grand Canyon.

Tarbuck/Lutgens, *Earth: An Introduction to Physical Geology, 8e*
© 2005 Pearson Prentice Hall, Inc.

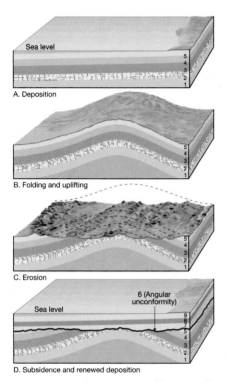

A. Deposition

B. Folding and uplifting

C. Erosion

D. Subsidence and renewed deposition

Figure 9.7 A,B,C,D Formation of an angular unconformity.

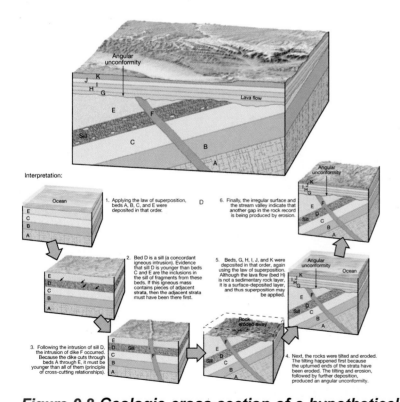

Figure 9.8 Geologic cross-section of a hypothetical region.

Tarbuck/Lutgens, *Earth: An Introduction to Physical Geology, 8e*

© 2005 Pearson Prentice Hall, Inc.

NOTES:

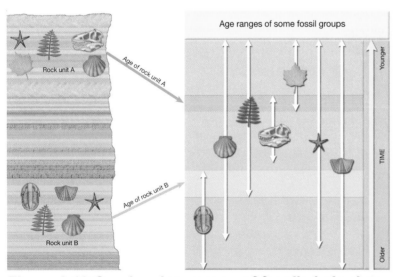

	Grand Canyon National Park	Zion National Park	Bryce Canyon National Park

Tertiary			Wasatch Fm
Cretaceous			Kaiparowits Fm / Wahweap Ss / Straight Cliffs Ss / Tropic Shale / Dakota Ss
Jurassic		Carmel Fm / Navajo Ss	Winsor Fm / Curtis Fm / Entrada Ss / Carmel Fm / Navajo Ss
Triassic	Moenkopi Fm	Kayenta Fm / Wingate Ss / Chinle Fm / Moenkopi Fm	Older rocks not exposed
Permian	Kaibab Ls / Toroweap Fm / Coconino Ss / Hermit Shale	Kaibab Ls / Older rocks not exposed	
Pennsylvanian	Supai Fm		
Mississippian	Redwall Ls		
Devonian	Temple Butte Ls / Muav Fm		
Cambrian	Bright Angel Shale / Tapeats Ss		
Precambrian	Colorado River / Vishnu Schist		

Photos by E.J. Tarbuck

Figure 9.9 Correlation of strata at three locations on the Colorado Plateau.

Age ranges of some fossil groups

Rock unit A

Age of rock unit A

Rock unit B

Age of rock unit B

Younger

TIME

Older

Figure 9.12 Overlapping ranges of fossils help date rocks.

Tarbuck/Lutgens, *Earth: An Introduction to Physical Geology, 8e*
© 2005 Pearson Prentice Hall, Inc.

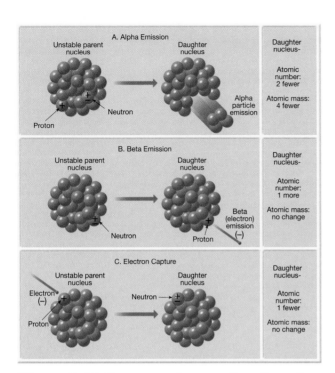

Figure 9.13 **Common types of radioactive decay.**

NOTES:

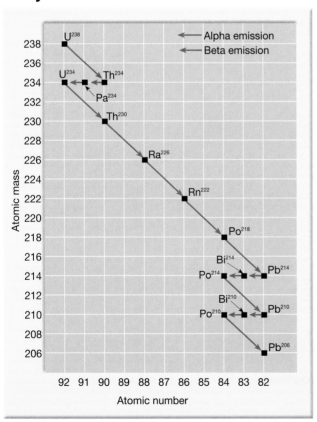

Figure 9.14 **U-238 radioactive decay series.**

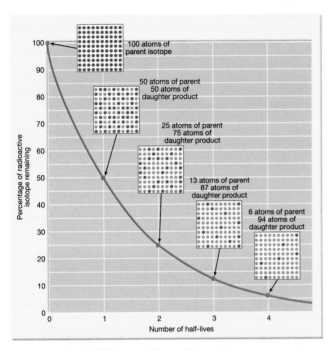

Figure 9.15 The radioactive decay curve.

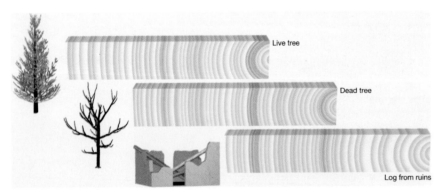

Figure 9.F Cross dating and dendrochronology.

NOTES:

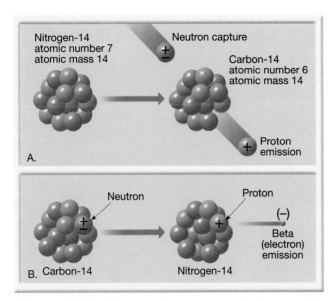

Figure 9.16 A,B *Production and decay of carbon-14.*

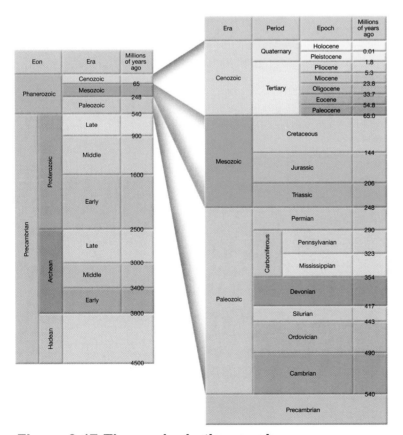

Figure 9.17 *The geologic time scale.*

Tarbuck/Lutgens, *Earth: An Introduction to Physical Geology, 8e*
© 2005 Pearson Prentice Hall, Inc.

NOTES:

Figure 9.H Chicxulub crater.

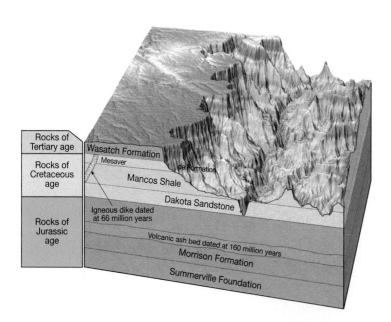

Figure 9.18 Numerical dates for sedimentary layers.

© 2005 Pearson Prentice Hall, Inc.

NOTES:

A. Undeformed strata (rock body)

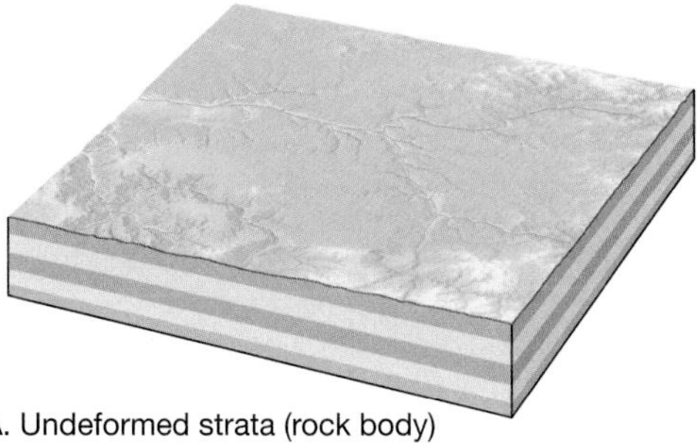

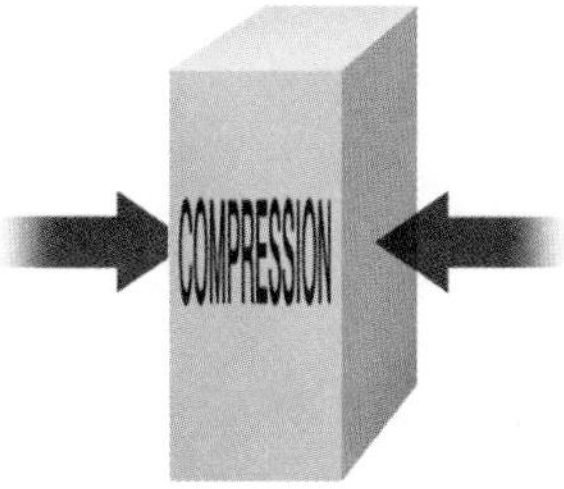

CUBE
OF
ROCK

B. Horizontal compressional stress causes rock bodies
to shorten horizontally and thicken vertically

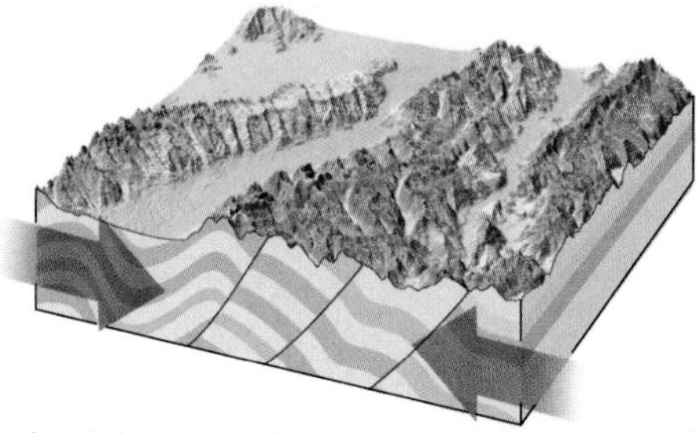

COMPRESSION

C. Horizontal tensional stress causes rock bodies
to lengthen horizontally and thin vertically

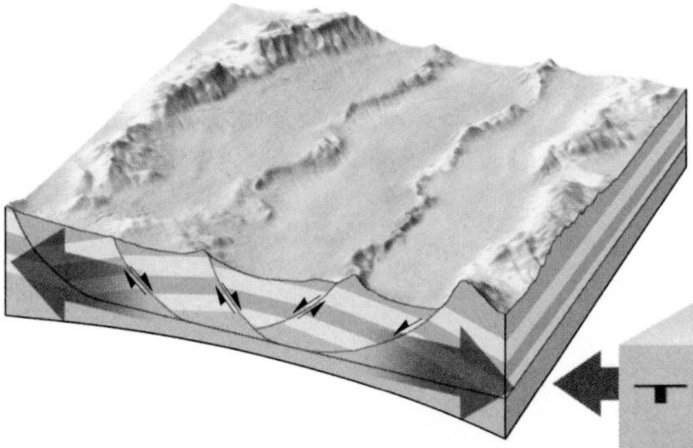

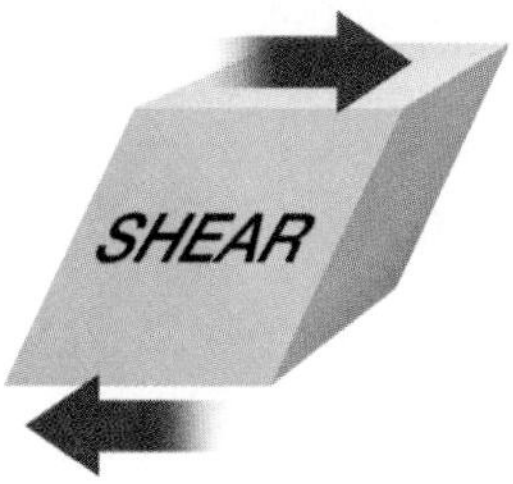

TENSION

D. Shear stress causes displacements
along fault zones or by ductile flow

SHEAR

**Figure 10.2 A,B,C,D Crustal deformation
caused by tectonic forces.**

NOTES:

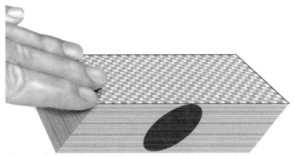

A. Deck of playing cards

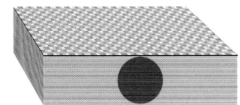

B. Shearing occurs when hand pushes top of deck

Figure 10.3 A,B *Illustration of shearing and resultant strain.*

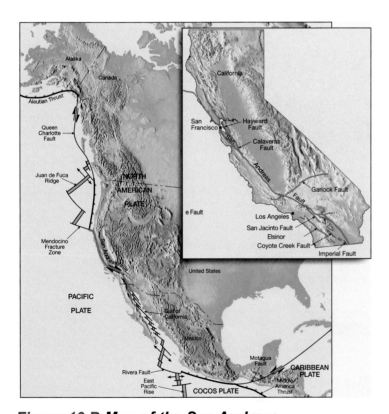

Figure 10.B *Map of the San Andreas fault system.*

NOTES:

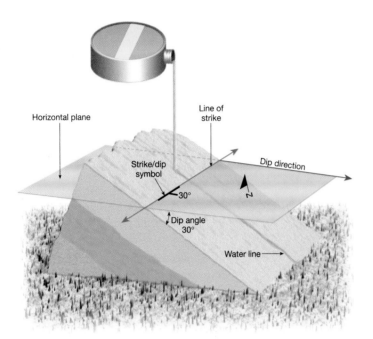

Figure 10.7 *Strike and dip of a rock layer.*

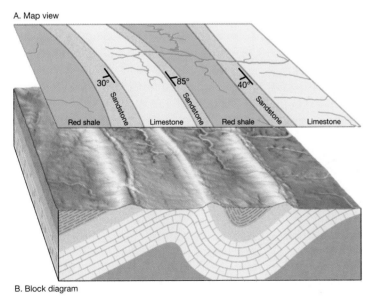

A. Map view

B. Block diagram

Figure 10.8 A,B *Strike and dip of outcropping sedimentary beds and the structure below.*

© 2005 Pearson Prentice Hall, Inc.

NOTES:

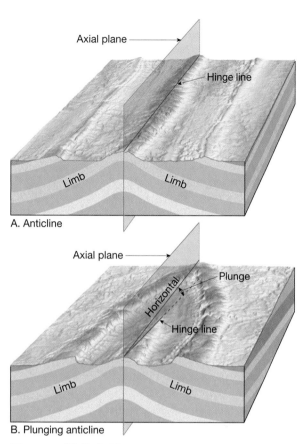

A. Anticline

B. Plunging anticline

Figure 10.9 A,B Features associated with symmetrical folds.

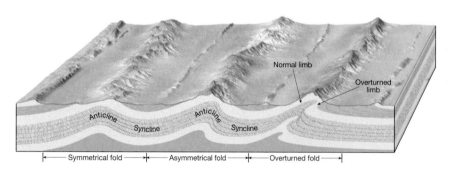

Figure 10.10 Principal types of folded strata.

© 2005 Pearson Prentice Hall, Inc.

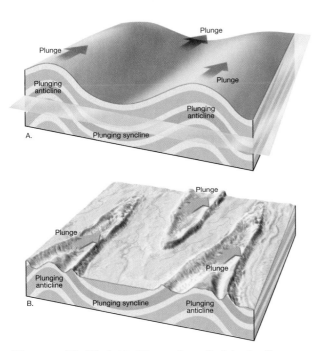

Figure 10.12 A,B Plunging folds before and after erosion.

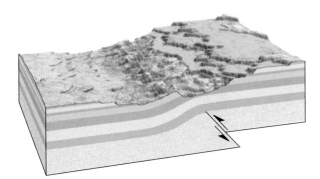

Figure 10.14 Monocline

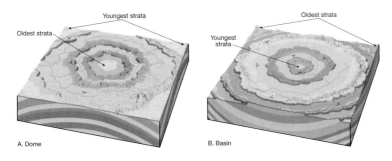

Figure 10.15 A,B Creation of domes and basins.

Tarbuck/Lutgens, *Earth: An Introduction to Physical Geology, 8e*
© 2005 Pearson Prentice Hall, Inc.

NOTES:

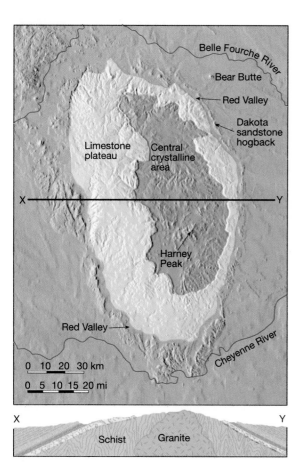

Figure 10.16 **The Black Hills of South Dakota.**

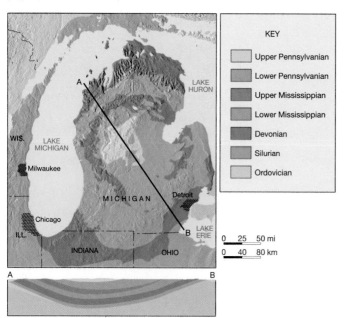

Figure 10.17 **Bedrock geology of the Michigan Basin.**

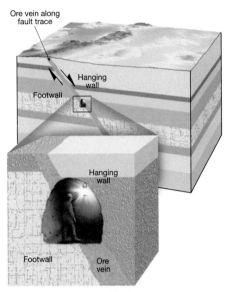

Figure 10.20 **Hanging wall and footwall.**

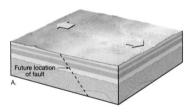

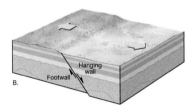

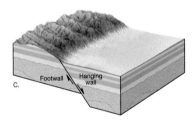

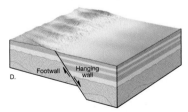

Figure 10.21 A,B,C,D **Block diagrams illlustrating a normal fault.**

© 2005 Pearson Prentice Hall, Inc.

NOTES:

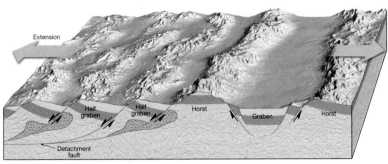

Figure 10.22 Normal faulting.

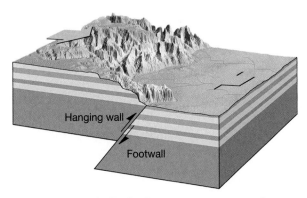

Figure 10.23 Relative movement along
a reverse fault.

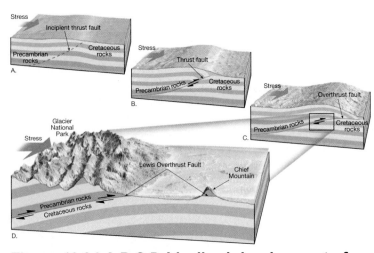

Figure 10.24 A,B,C,D Idealized development of
Lewis Overthrust fault.

NOTES:

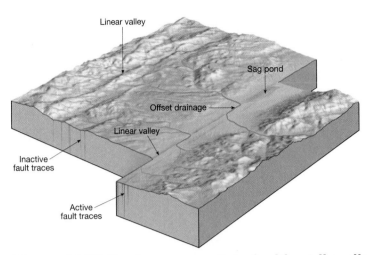

Figure 10.27 Features associated with strike-slip faults.

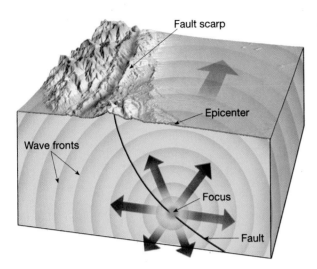

Figure 11.2 Earthquake focus and epicenter.

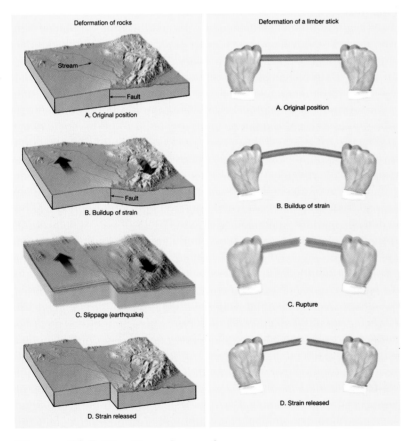

Figure 11.5 Elastic rebound.

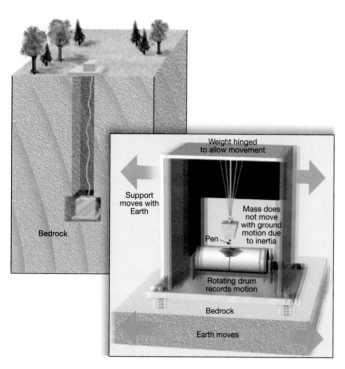

*Figure 11.7 **Principle of the seismograph.***

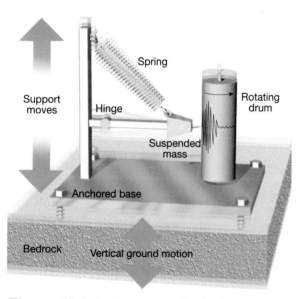

*Figure 11.8 **Seismograph designed to record vertical ground motion.***

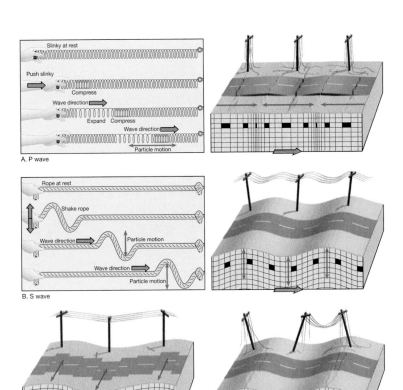

Figure 11.9 A,B,C,D Types of seismic waves and their characteristic motions.

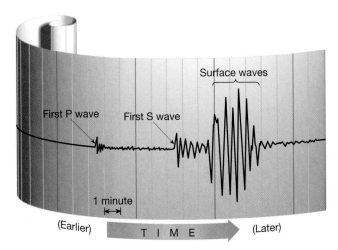

Figure 11.10 Typical seismogram.

NOTES:

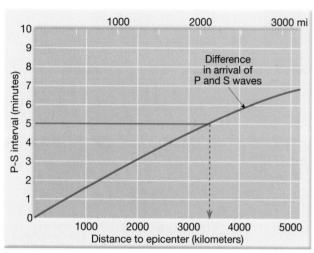

Figure 11.11 *A travel-time graph.*

Figure 11.12 *Location of earthquake epicenter.*

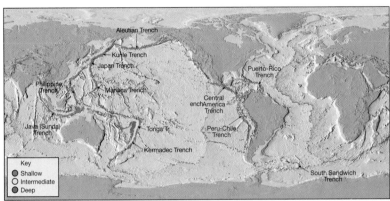

Figure 11.13 *World earthquake distribution.*

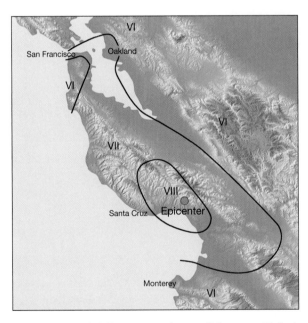

Figure 11.14 *Destruction of Loma Prieta earthquake in 1989.*

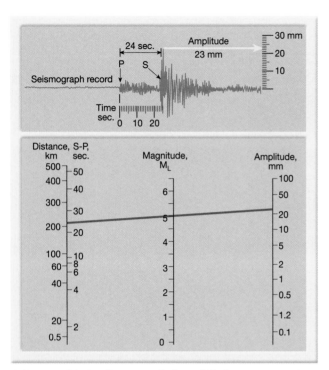

Figure 11.15 Determining Richter magnitude.

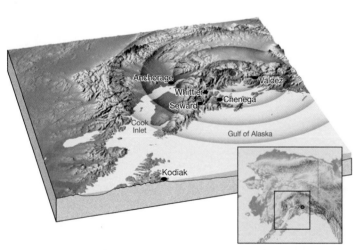

Figure 11.16 Region most affected by Alaska
earthquake of 1964.

© 2005 Pearson Prentice Hall, Inc.

NOTES:

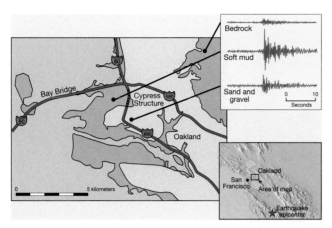

Figure 11.C Loma Prieta earthquake damage.

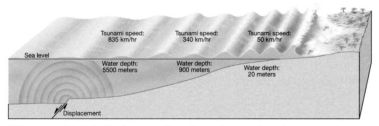

Figure 11.20 Schematic drawing of a tsunami.

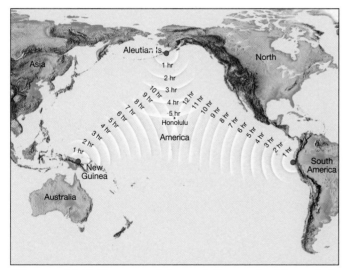

Figure 11.D Tsunami travel times to Honolulu, Hawaii.

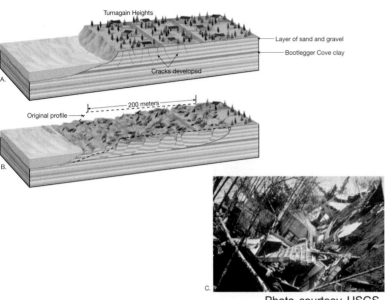

Figure 11.22 A,B,C Turnagain Heights slide: 1964 Alaskan earthquake.

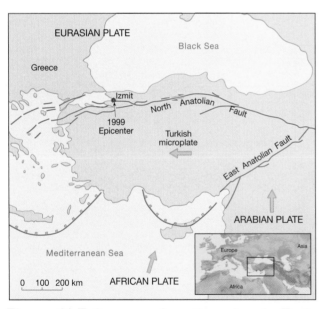

Figure 11.E Causes of earthquakes in Turkey.

NOTES:

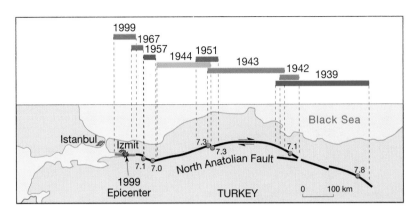

Figure 11.F North Anotolian earthquakes, 1939-1999.

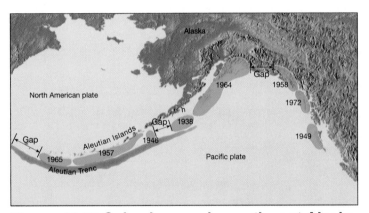

Figure 11.25 Seismic gaps in southwest Alaska.

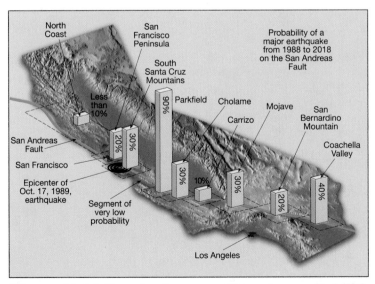

Figure 11.26 San Andreas earthquake probabilities, 1988-2018.

Tarbuck/Lutgens, *Earth: An Introduction to Physical Geology, 8e*
© 2005 Pearson Prentice Hall, Inc.

NOTES:

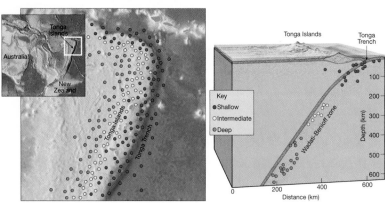

Figure 11.27 *Distribution of earthquake foci near Tonga, 1965.*

NOTES:

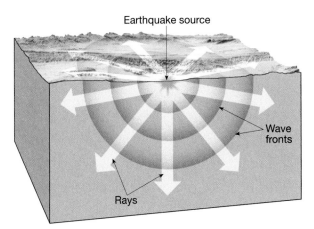

Figure 12.1 Seismic energy travels in all directions.

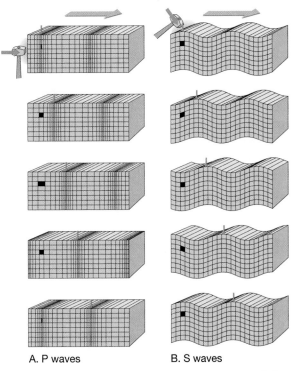

A. P waves   B. S waves

Figure 12.2 A,B The transmission of P and S waves through a solid.

NOTES:

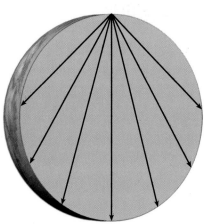

Figure 12.3 Seismic waves in a uniform planet.

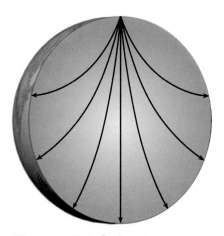

Figure 12.4 Seismic waves: velocity increases with depth.

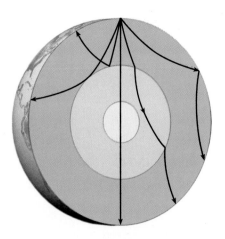

Figure 12.5 Seismic wave paths through Earth.

Tarbuck/Lutgens, *Earth: An Introduction to Physical Geology, 8e*
© 2005 Pearson Prentice Hall, Inc.

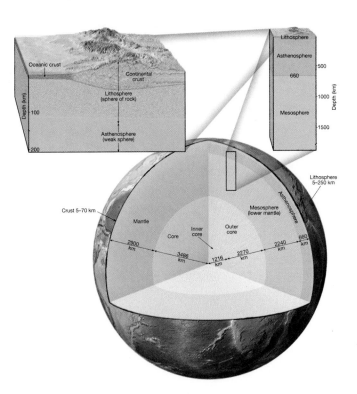

Figure 12.6 Views of Earth's layered structure.

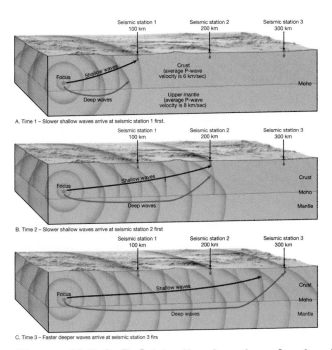

A. Time 1 – Slower shallow waves arrive at seismic station 1 first.

B. Time 2 – Slower shallow waves arrive at seismic station 2 first

C. Time 3 – Faster deeper waves arrive at seismic station 3 firs

Figure 12.7 A, B,C Idealized paths of seismic waves.

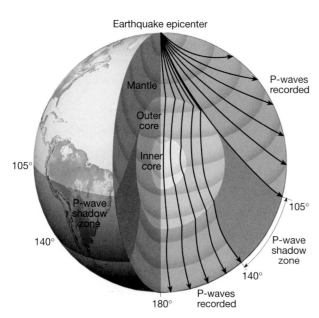

Figure 12.8 Seismic waves and the core-mantle boundary.

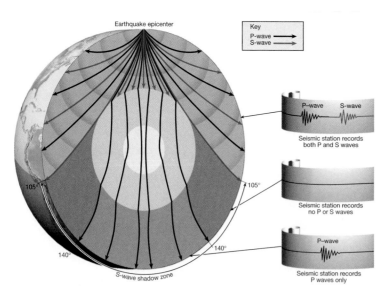

Figure 12.9 View of Earth's interior showing P and S wave paths.

Tarbuck/Lutgens, *Earth: An Introduction to Physical Geology, 8e*
© 2005 Pearson Prentice Hall, Inc.

NOTES:

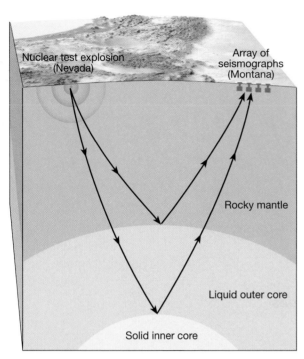

Figure 12.10 *Travel times of seismic waves.*

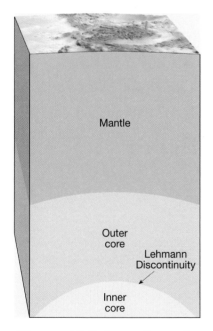

Figure 12.B *Location of Lehmann discontinuity.*

NOTES:

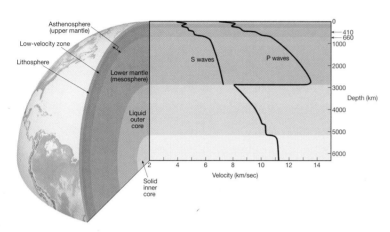

Figure 12.11 *Variations in P and S wave velocities with depth.*

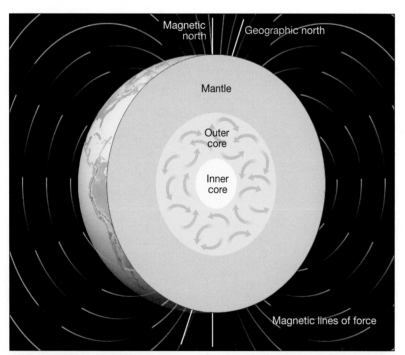

Figure 12.C *Earth's magnetic field and the outer core.*

Tarbuck/Lutgens, *Earth: An Introduction to Physical Geology, 8e*
© 2005 Pearson Prentice Hall, Inc.

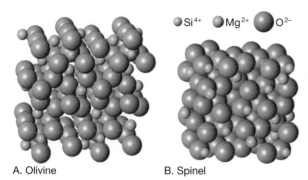

A. Olivine   B. Spinel

Figure 12.12 A,B Comparison of olivine and spinel.

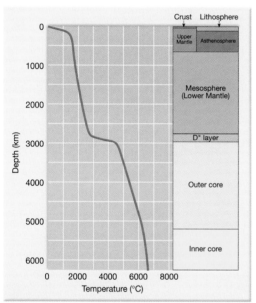

Figure 12.13 Estimated geothermal gradient for Earth.

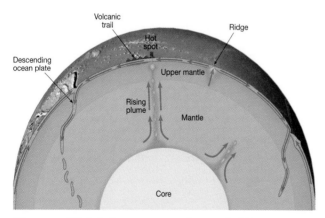

Figure 12.14 A proposed model for convective flow in the mantle.

Tarbuck/Lutgens, *Earth: An Introduction to Physical Geology, 8e*
© 2005 Pearson Prentice Hall, Inc.

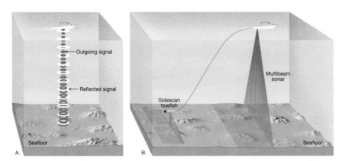

Figure 13.2 A,B Echo sounder and multibeam sonar.

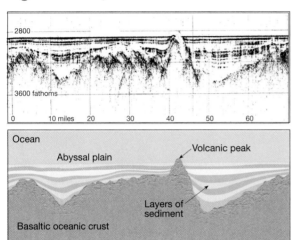

Figure 13.4 Seismic cross section and matching sketch.

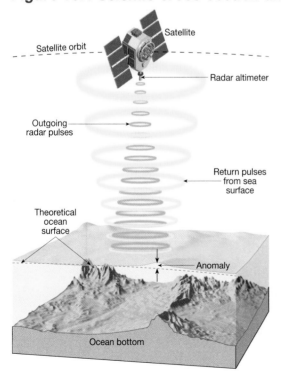

Figure 13.5 Satellite altimeter.

Tarbuck/Lutgens, *Earth: An Introduction to Physical Geology, 8e*
© 2005 Pearson Prentice Hall, Inc.

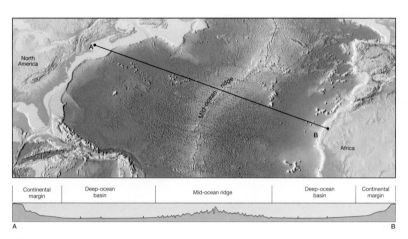

Figure 13.6 *Major topographic divisions of the North Atlantic.*

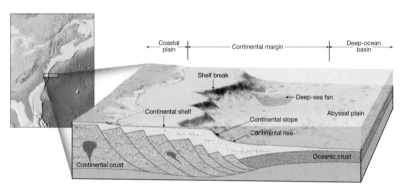

Figure 13.7 *The provinces of a passive continental margin.*

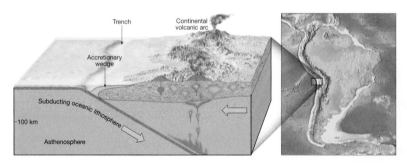

Figure 13.8 *Active continental margin.*

Tarbuck/Lutgens, *Earth: An Introduction to Physical Geology, 8e*
© 2005 Pearson Prentice Hall, Inc.

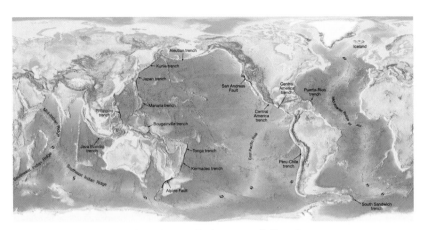

Figure 13.9 **Distribution of the world's deep ocean trenches.**

Figure 13.11 A,B,C **Distribution of the oceanic ridge system.**

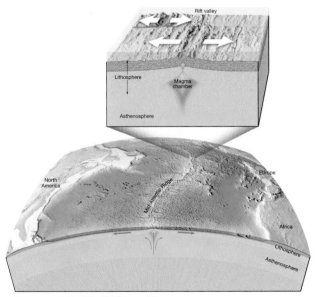

Figure 13.12 **Rift valleys.**

NOTES:

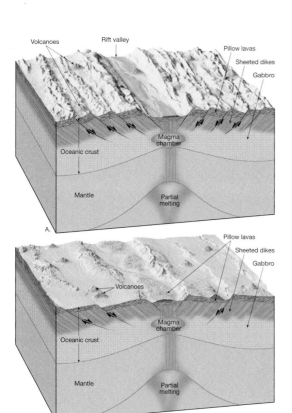

Figure 13.13 A,B Topography of the crest of an oceanic ridge.

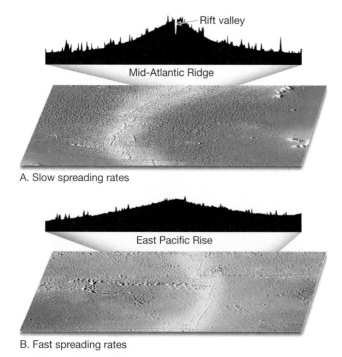

A. Slow spreading rates

B. Fast spreading rates

Figure 13.14 A,B Slow and fast spreading rates.

Tarbuck/Lutgens, *Earth: An Introduction to Physical Geology, 8e*

© 2005 Pearson Prentice Hall, Inc.

NOTES:

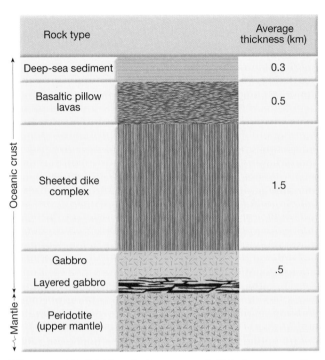

Figure 13.15 *Rock type and thickness of a typical section of oceanic crust.*

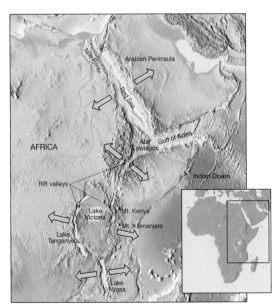

Figure 13.18 *East African rift valleys.*

NOTES:

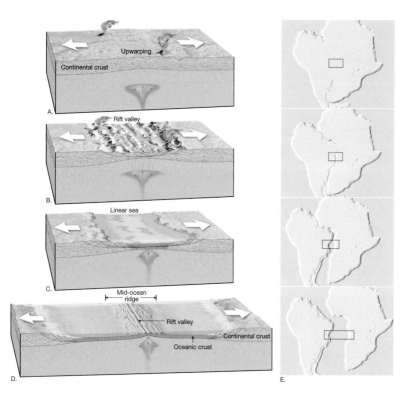

Figure 13.19 A,B,C,D,E *Formation of an ocean basin.*

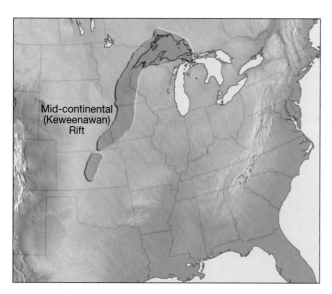

Figure 13.20 *Location of a failed rift.*

NOTES:

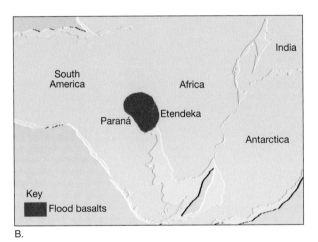

Figure 13.21 A,B Evidence for the role of mantle plumes in rifting.

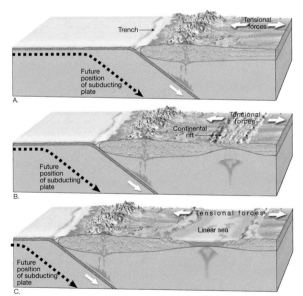

Figure 13.22 A,B,C Trench retreat, or roll back.

NOTES:

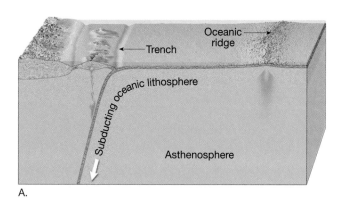

A.

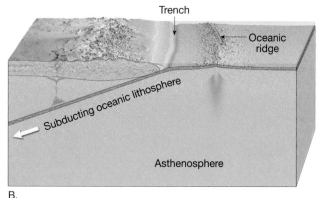

B.

Figure 13.23 A,B *The angle at which oceanic lithosphere descends into the asthenosphere depends on its density.*

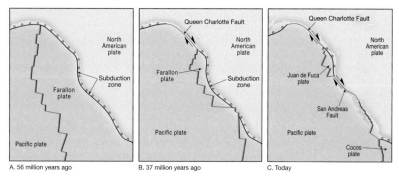

Figure 13.24 A,B,C *Demise of the Farallon plate.*

Tarbuck/Lutgens, *Earth: An Introduction to Physical Geology, 8e*
© 2005 Pearson Prentice Hall, Inc.

NOTES:

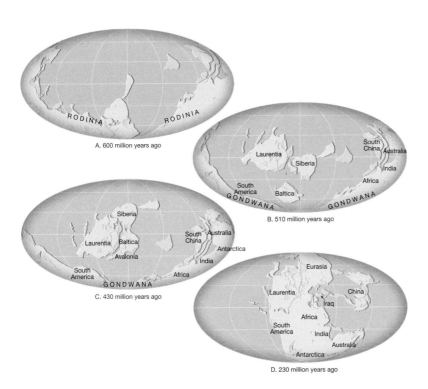

Figure 13.25 A,B,C,D Breakup and dispersal of Rodinia and formation of Pangaea.

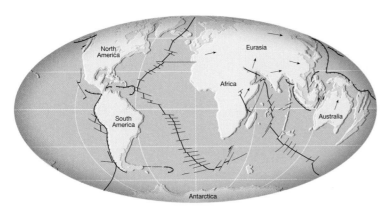

Figure 13.26 The world as it may look 50 million years from now.

© 2005 Pearson Prentice Hall, Inc.

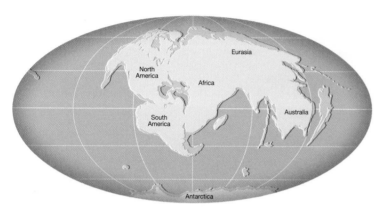

Figure 13.27 The world as it may look 250 million years from now.

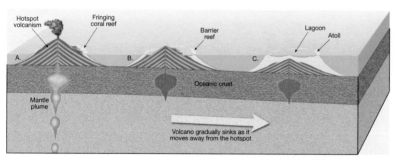

Figure 13.C A,B,C, Formation of a coral atoll.

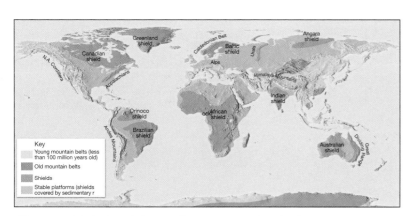

Figure 14.3 Earth's major mountain belts.

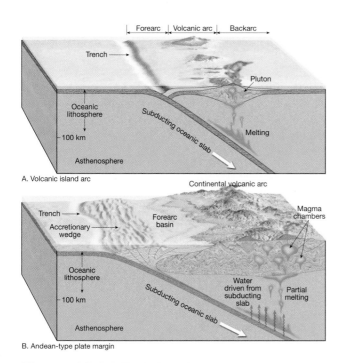

Figure 14.4 A,B Volcanic island arc and an Andean-type plate margin.

NOTES:

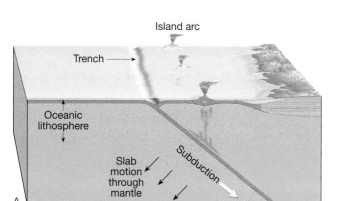

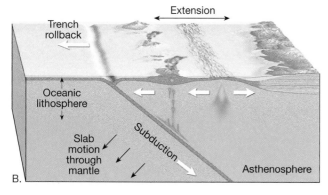

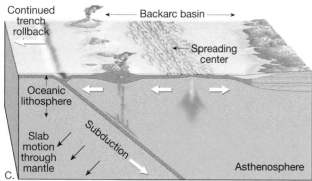

Figure 14.6 A,B,C Formation of a backarc basin.

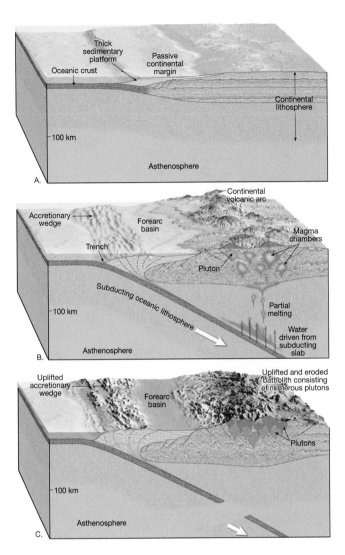

*Figure 14.7 A,B,C Orogenesis along an
Andean-type subduction zone.*

© 2005 Pearson Prentice Hall, Inc.

NOTES:

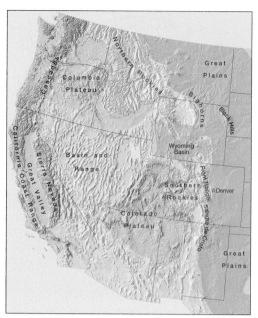

Figure 14.8 Mountainous landforms of the western U.S.

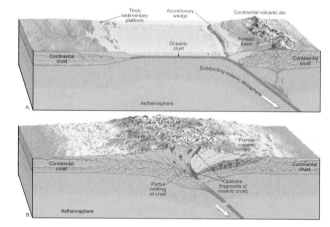

Figure 14.9 A,B Formation of a compressional mountain belt.

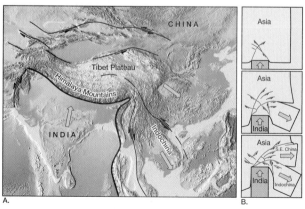

Figure 14.11 A,B Collision between India and Asia.

Tarbuck/Lutgens, *Earth: An Introduction to Physical Geology, 8e*

© 2005 Pearson Prentice Hall, Inc.

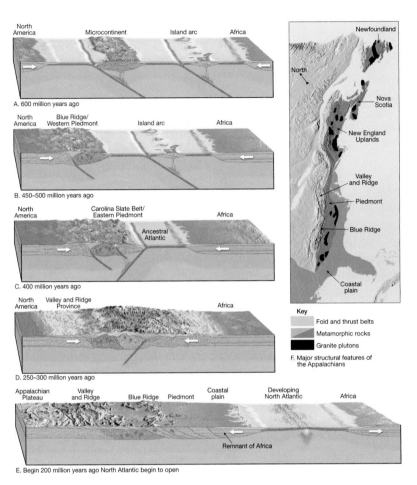

Figure 14.12 Development of the southern Appalachians.

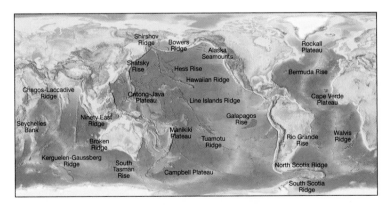

Figure 14.14 Distribution of present-day oceanic plateaus and other submerged crustal fragments.

© 2005 Pearson Prentice Hall, Inc.

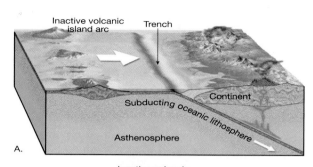

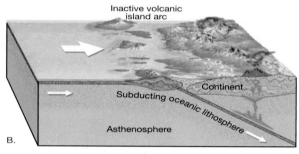

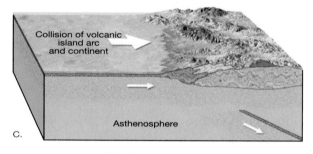

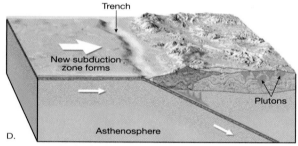

Figure 14.15 A,B,C,D Accretion of an inactive volcanic island arc.

Tarbuck/Lutgens, *Earth: An Introduction to Physical Geology, 8e*
© 2005 Pearson Prentice Hall, Inc.

NOTES:

NOTES:

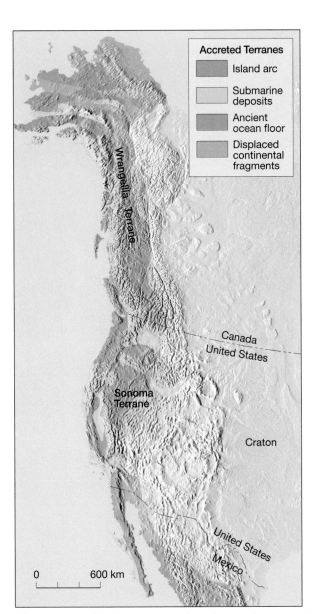

Figure 14.16 Terranes that have been added to western North America during the past 200 million years.

NOTES:

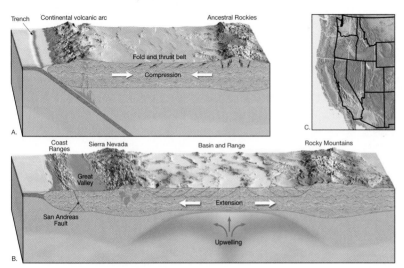

Figure 14.18 A,B Basin and Range province.

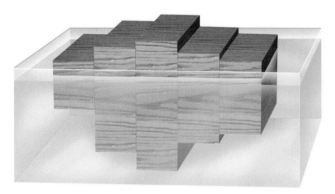

Figure 14.20 Demonstration of isostasy.

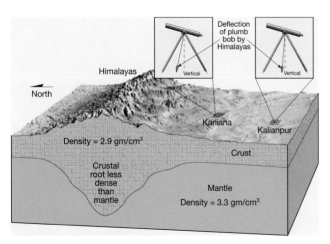

Figure 14.D Himalayas gravity.

Tarbuck/Lutgens, *Earth: An Introduction to Physical Geology, 8e*
© 2005 Pearson Prentice Hall, Inc.

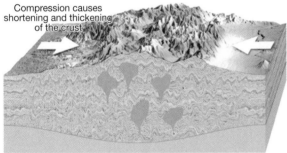

Figure 14.21 A,B,C *Erosion and isostatic adjustment.*

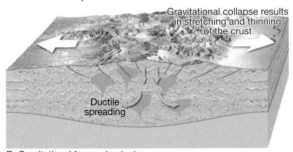

A. Horizontal compressional forces dominate

B. Gravitational forces dominate

Figure 14.22 A,B *Block diagram of a mountain belt that is collapsing under its own "weight."*

Tarbuck/Lutgens, *Earth: An Introduction to Physical Geology, 8e*
© 2005 Pearson Prentice Hall, Inc.

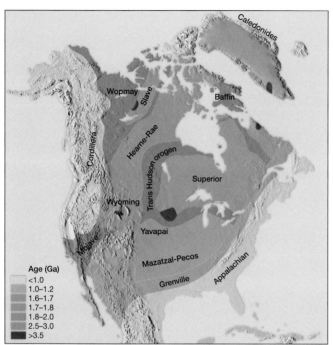

Figure 14.23 Major Precambrian mountain belts and cores of ancient continental blocks and their ages.

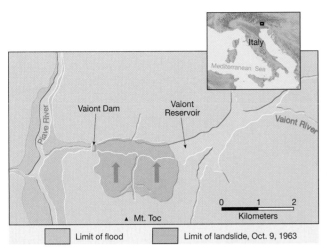

Figure 15.A Sketch map of the Vaiont River area.

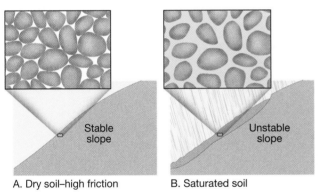

Figure 15.4 A,B The effect of water on mass wasting.

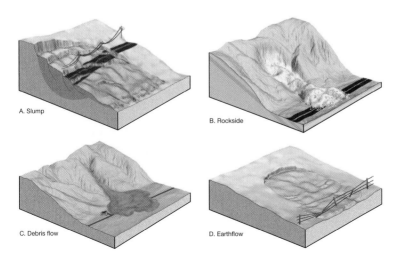

Figure 15.7 A,B,C,D Four rapid forms of mass wasting.

© 2005 Pearson Prentice Hall, Inc.

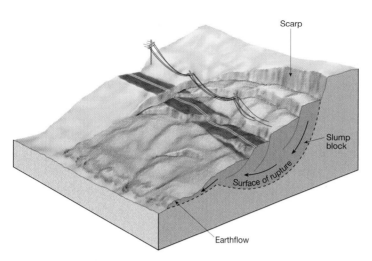

Figure 15.11 **Slump.**

NOTES:

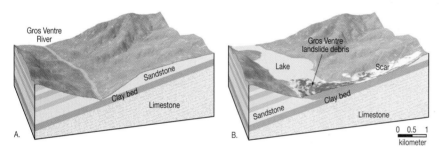

Figure 15.14 A,B **Gros Ventre rockslide: before and after.**

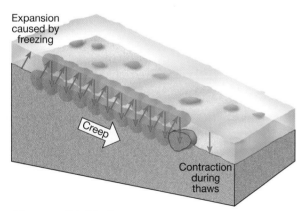

Figure 15.18 **Creep.**

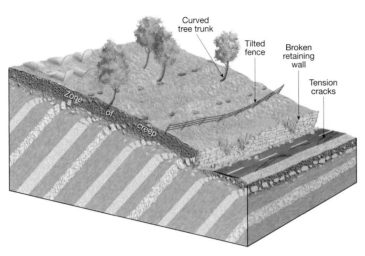

Figure 15.19 *Effects of creep.*

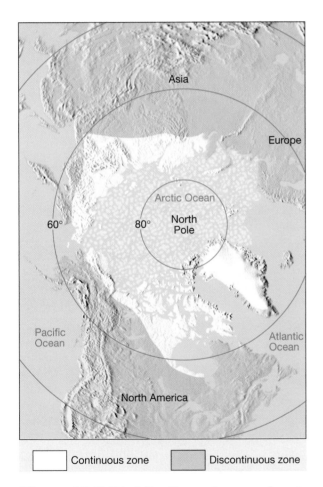

Figure 15.G *Distribution of permafrost in the Northern Hemisphere.*

Tarbuck/Lutgens, *Earth: An Introduction to Physical Geology, 8e*
© 2005 Pearson Prentice Hall, Inc.

NOTES:

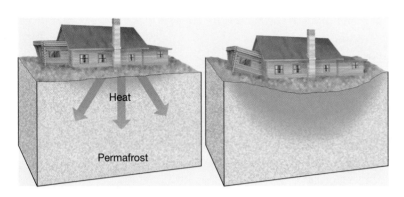

Figure 15.I Subsidence caused by thawing permafrost.

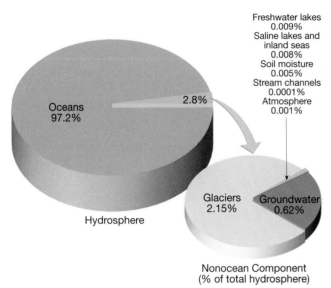

Figure 16.2 Distribution of Earth's water.

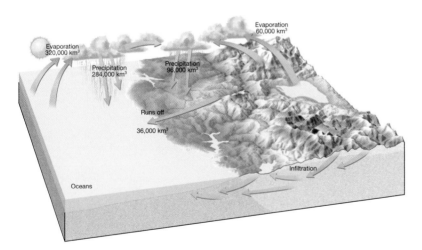

Figure 16.3 Earth's water balance.

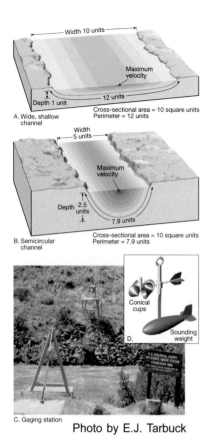

NOTES:

Photo by E.J. Tarbuck

Figure 16.5 A,B,C,D Influence of channel shape on velocity.

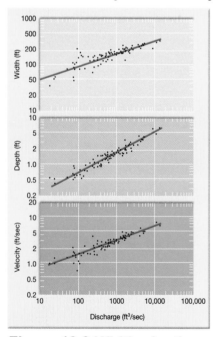

Figure 16.6 Width, depth, velocity of the Powder River at Locate, Montana.

NOTES:

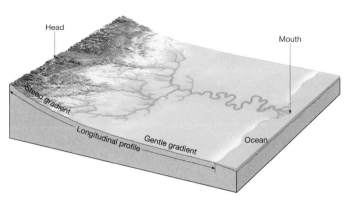

Figure 16.7 A longitudinal profile of a stream.

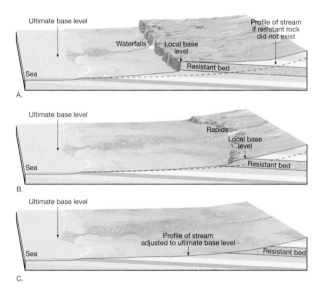

Figure 16.8 A,B,C A resistant layer of rock can act as a local base level.

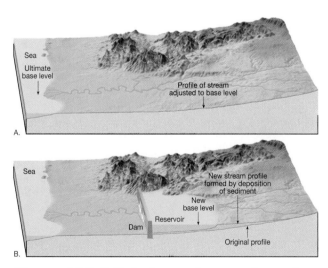

Figure 16.9 A,B A dam raises a stream's base level.

Tarbuck/Lutgens, *Earth: An Introduction to Physical Geology, 8e*
© 2005 Pearson Prentice Hall, Inc.

NOTES:

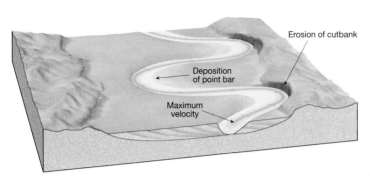

Figure 16.14 Stream meander and point bars.

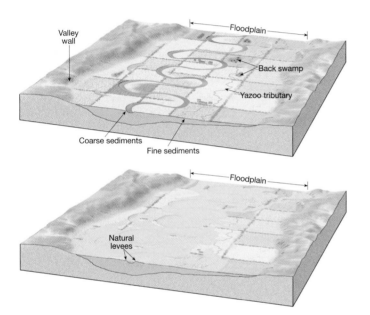

Figure 16.16 Natural levees are gently sloping structures created by repeated floods.

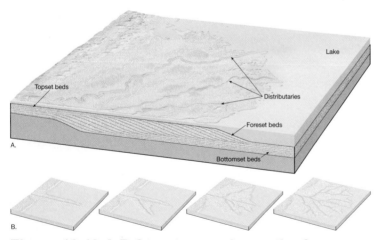

Figure 16.18 A,B Structure and growth of a simple delta.

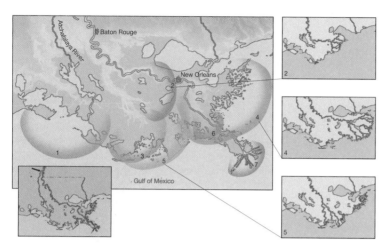

Figure 16.20 Evolution of the Mississippi River delta.

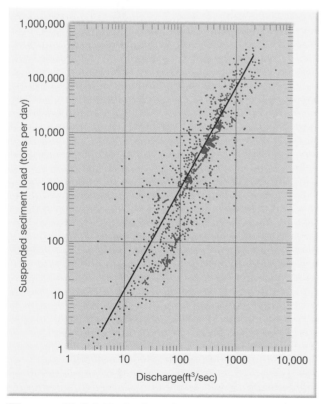

Figure 16.21 Relationship between suspended load and discharge.

© 2005 Pearson Prentice Hall, Inc.

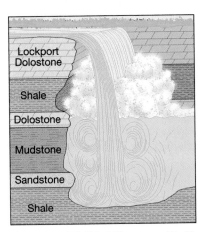

Figure 16.23 *Niagara Falls geology.*

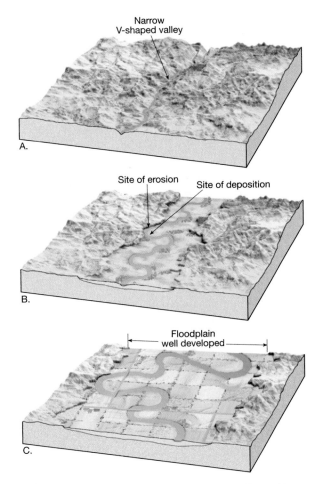

Figure 16.24 A,B,C *Stream eroding its floodplain.*

NOTES:

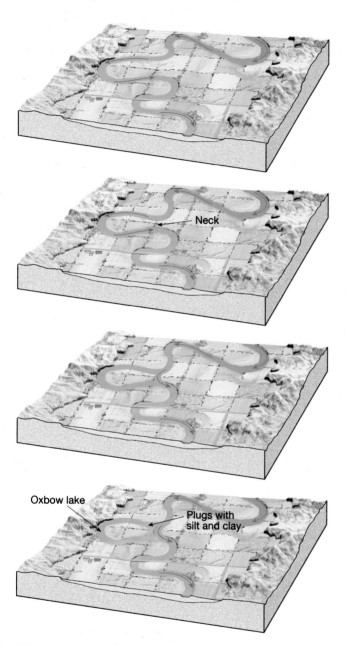

Figure 16.26 **Formation of a cutoff and oxbow lake.**

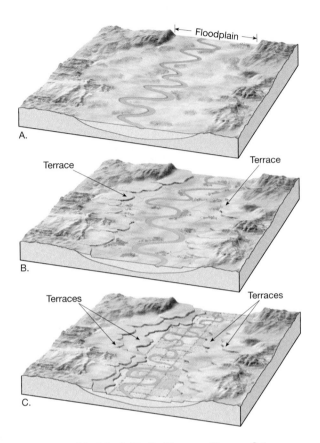

Figure 16.29 A,B,C **Formation of terraces.**

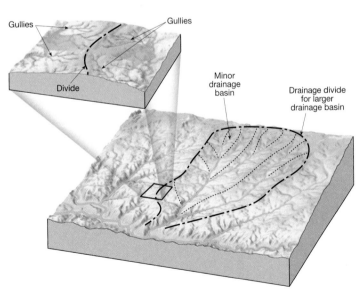

Figure 16.30 **A drainage basin and a divide.**

© 2005 Pearson Prentice Hall, Inc.

Figure 16.31 *The drainage basin of the Mississippi River.*

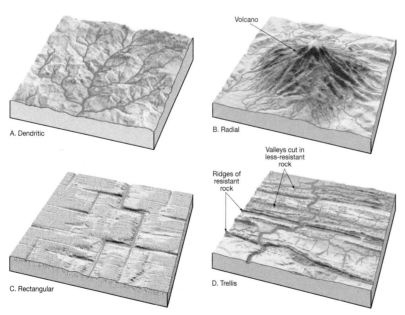

Figure 16.32 *Drainage patterns: A. Dendritic. B. Radial. C. Rectangular. D. Trellis.*

NOTES:

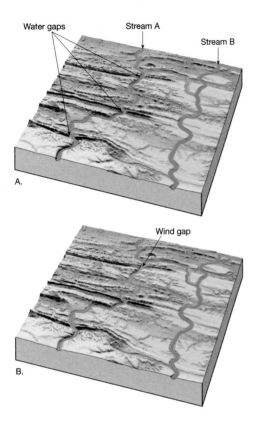

Figure 16.34 A,B Stream piracy and the formation of wind gaps.

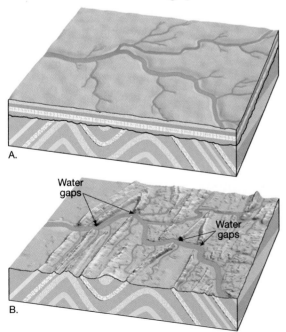

Figure 16.35 A,B Development of a superposed stream.

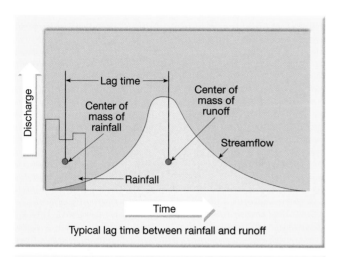

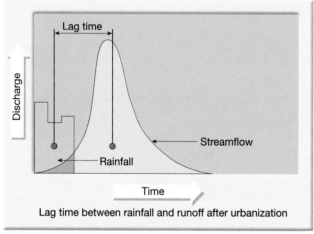

Figure 16.37 Lag time between rainfall and runoff.

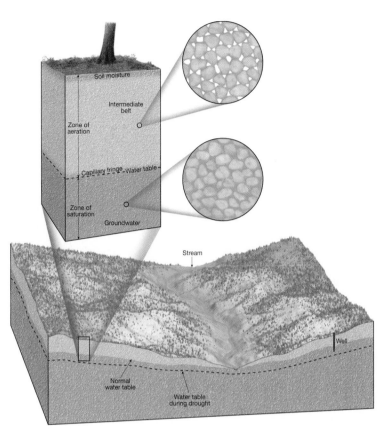

Figure 17.2 Distribution of underground water.

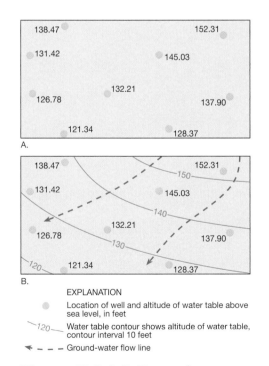

EXPLANATION

Location of well and altitude of water table above sea level, in feet

120 — Water table contour shows altitude of water table, contour interval 10 feet

— — — Ground-water flow line

Figure 17.3 A,B Preparing a map of the water table.

Tarbuck/Lutgens, *Earth: An Introduction to Physical Geology, 8e*
© 2005 Pearson Prentice Hall, Inc.

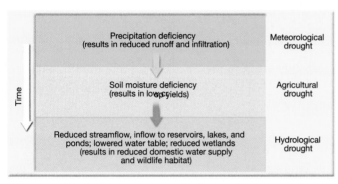

Figure 17.A **Sequence of drought impacts.**

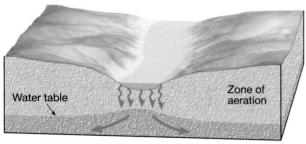

A. Gaining stream

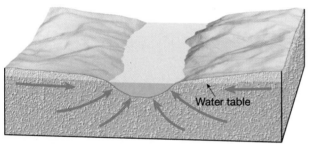

B. Losing stream (connected)

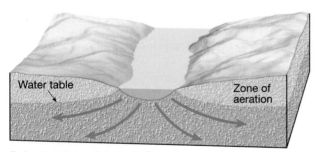

C. Losing stream (disconnected)

Figure 17.4 A,B,C **Interaction between the groundwater system and streams.**

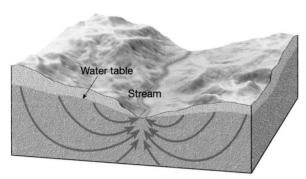

Figure 17.5 **Groundwater movement.**

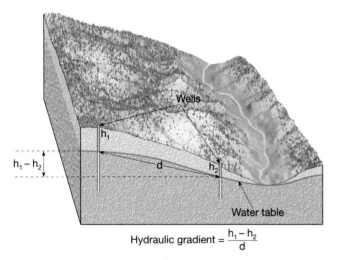

Hydraulic gradient = $\dfrac{h_1 - h_2}{d}$

Figure 17.6 **Hydraulic gradient.**

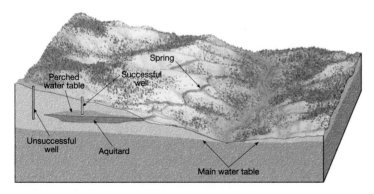

Figure 17.8 **Aquitard and perched water table.**

Tarbuck/Lutgens, *Earth: An Introduction to Physical Geology, 8e*
© 2005 Pearson Prentice Hall, Inc.

NOTES:

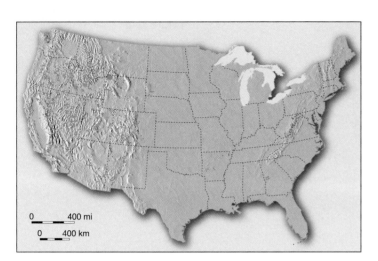

Figure 17.9 Distribution of hot springs and geysers in the U.S.

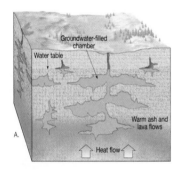

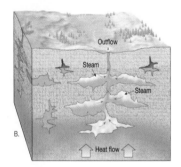

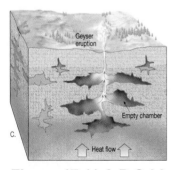

Figure 17.11 A,B,C Idealized diagrams of a geyser.

Tarbuck/Lutgens, *Earth: An Introduction to Physical Geology, 8e*

© 2005 Pearson Prentice Hall, Inc.

NOTES:

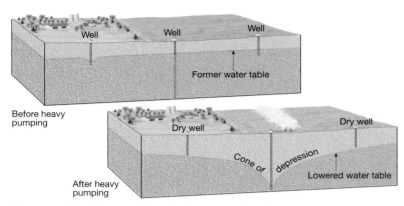

Figure 17.13 *A cone of depression in the water table.*

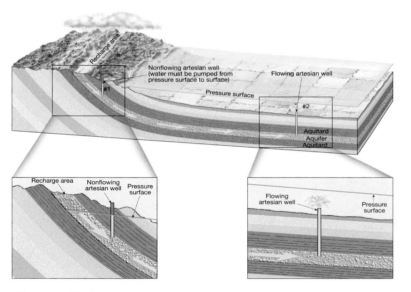

Figure 17.15 *Artesian systems.*

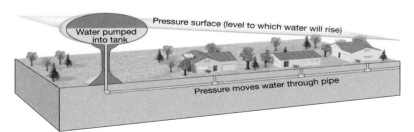

Figure 17.17 *City water systems are artificial artesian systems.*

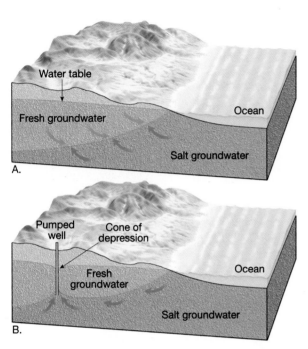

Figure 17.19 A,B *Saltwater contamina-tion of wells.*

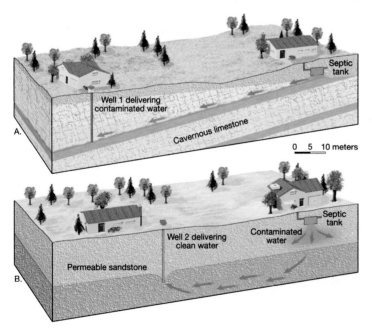

Figure 17.20 A,B *Contaminated groundwater and wells.*

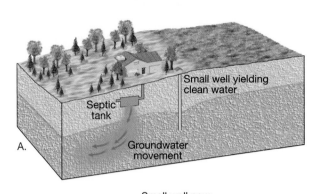

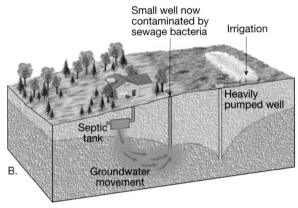

Figure 17.21 A,B **Septic tank and ground-water contamination.**

NOTES:

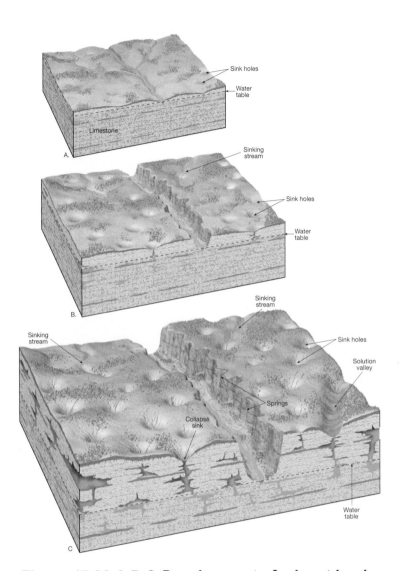

Figure 17.26 A,B,C Development of a karst landscape.

NOTES:

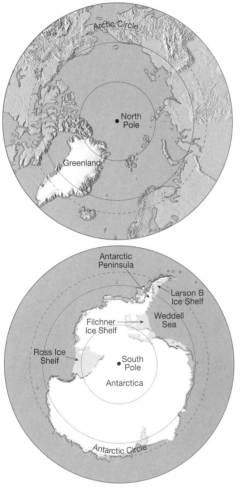

Figure 18.3 *Continental ice sheets: Greenland and Antarctica.*

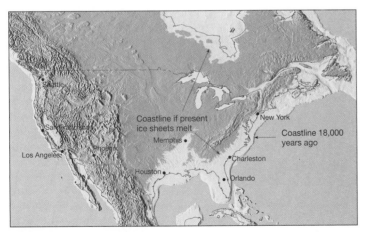

Figure 18.6 *Comparison of coastlines under varying amounts of glaciation.*

© 2005 Pearson Prentice Hall, Inc.

NOTES:

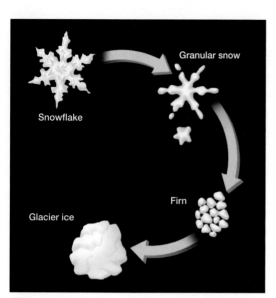

Figure 18.7 Freshly fallen snow changing to dense glacial ice.

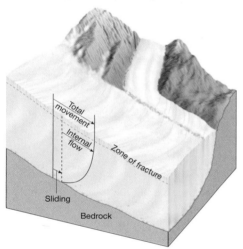

Figure 18.8 Vertical cross-section through a glacier.

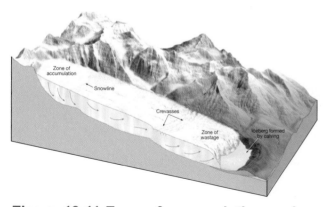

Figure 18.11 Zone of accumulation and zone of wastage.

Tarbuck/Lutgens, *Earth: An Introduction to Physical Geology, 8e*

© 2005 Pearson Prentice Hall, Inc.

NOTES:

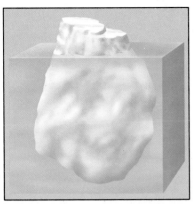

*Figure 18.12 **Only about 20 percent of an iceberg protrudes above the water line.***

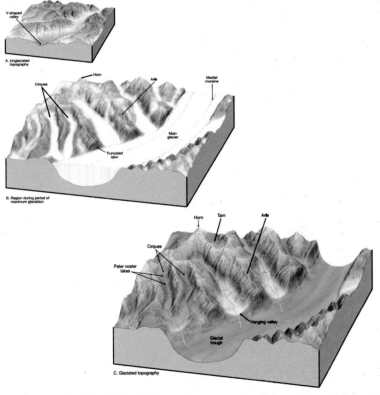

*Figure 18.15 A,B,C **Development of erosional landforms created by alpine glaciers.***

NOTES:

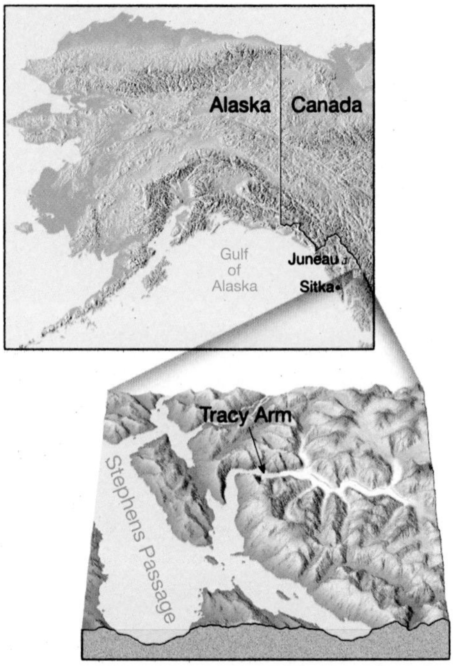

Figure 18.17 **The fjord at Tracy Arm, Alaska is a drowned glacial trough.**

Photo by E.J. Tarbuck

Figure 18.18 **Roche Moutonnée, in Yosemite National Park.**

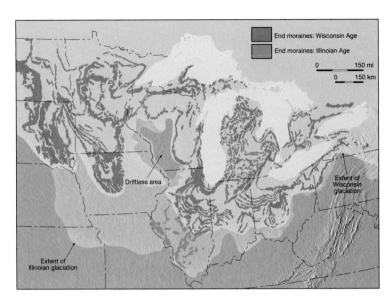

Figure 18.23 **End moraines of the Great Lakes region.**

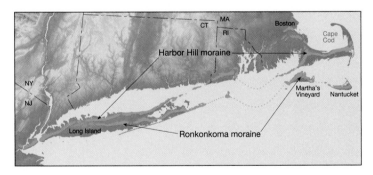

Figure 18.24 **End moraines along the U.S. East coast.**

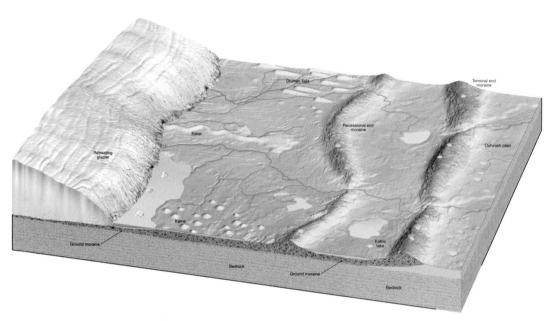

Figure 18.25 *Depositional landforms.*

NOTES:

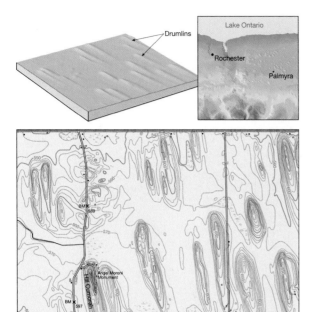

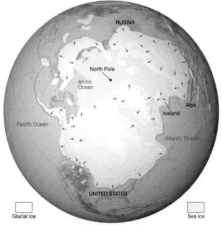

Figure 18.26 Drumlin field on a topographic map.

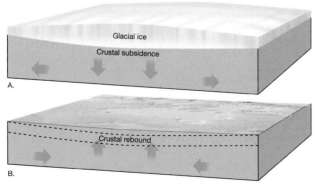

Figure 18.27 Maximum extent of ice sheets during the Ice Age.

Figure 18.28 A,B Crustal subsidence and rebound.

Tarbuck/Lutgens, *Earth: An Introduction to Physical Geology, 8e*
© 2005 Pearson Prentice Hall, Inc.

NOTES:

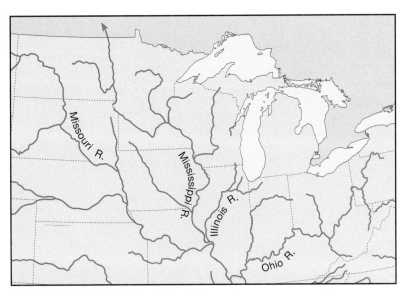

Figure 18.B Central United States after the Ice Age.

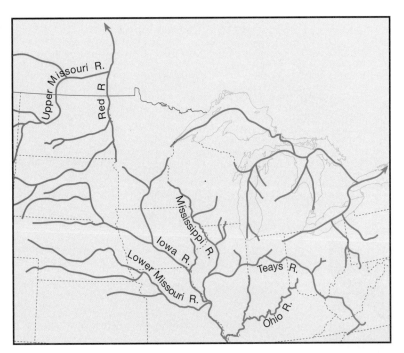

Figure 18.C Central United States prior to the Ice Age.

© 2005 Pearson Prentice Hall, Inc.

NOTES:

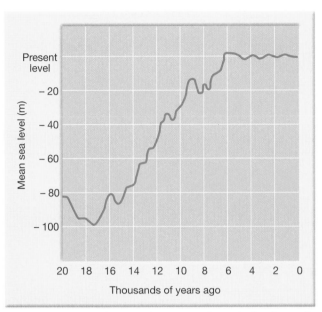

Figure 18.29 Changing sea level during the last 20,000 years.

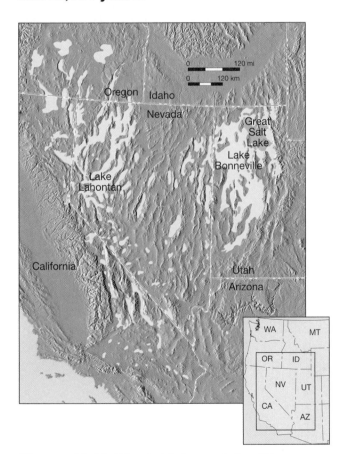

Figure 18.30 Pluvial lakes of the Western United States.

NOTES:

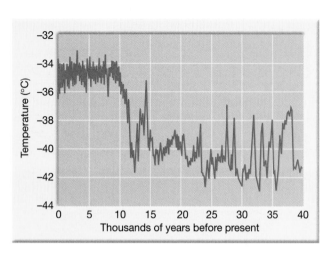

Figure 18.E Temperature variations over the last 40,000 years.

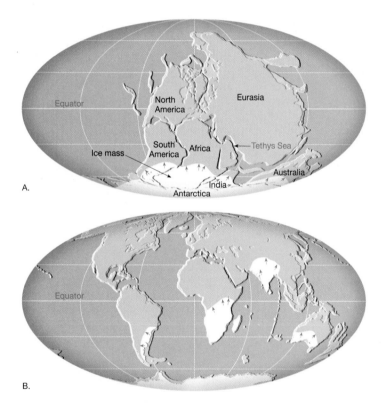

Figure 18.31 A,B Glaciation 300 million years ago and today.

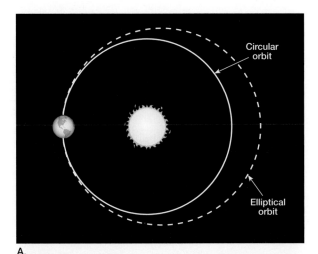

A.

NOTES:

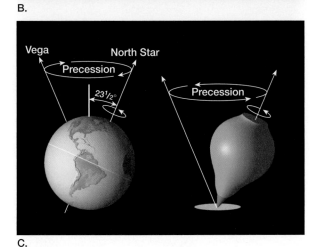

B.

C.

Figure 18.32 A,B,C Orbital variations of Earth.

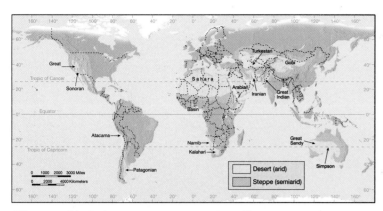

Figure 19.2 *Arid and semiarid climates.*

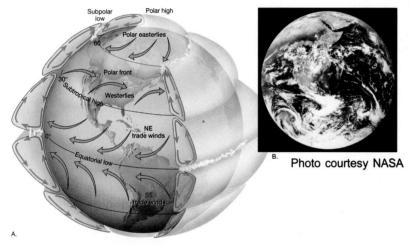

B.   Photo courtesy NASA

A.

Figure 19.3 A,B *Idealized diagram of Earth's general circulation.*

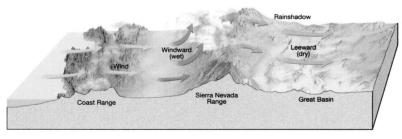

Figure 19.4 *The Great Basin is a rainshadow desert.*

NOTES:

Figure 19.A *The Aral Sea.*

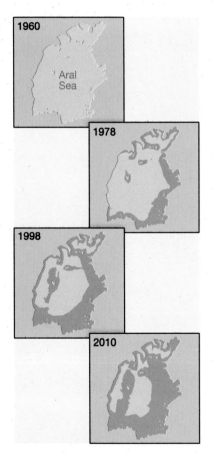

Figure 19.B *The shrinking Aral Sea.*

NOTES:

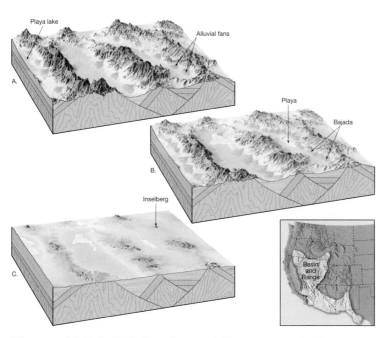

Figure 19.7 A,B,C Basin and Range evolution.

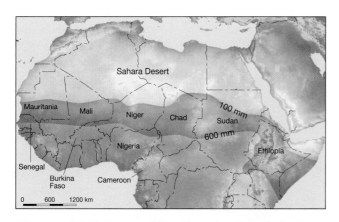

Figure 19.E Desertification in the Sahel region of the Sahara.

© 2005 Pearson Prentice Hall, Inc.

NOTES:

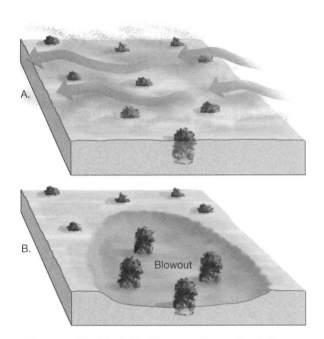

Figure 19.12 A,B *Formation of a blowout.*

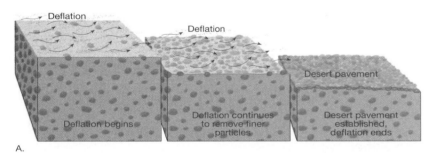

Figure 19.13 A *Formation of desert pavement.*

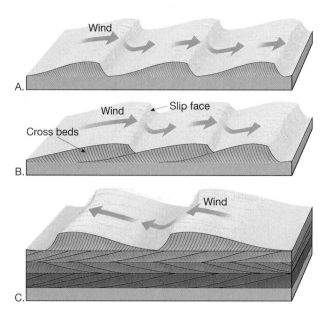

Figure 19.16 A,B,C Dunes commonly have an asymmetrical shape.

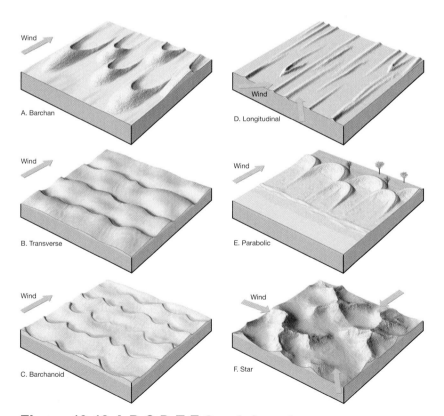

Figure 19.18 A,B,C,D,E,F Sand dune types.

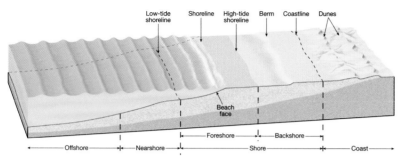

Figure 20.2 **The coastal zone.**

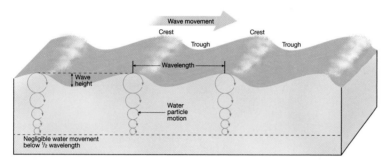

Figure 20.3 **The basic parts of a wave.**

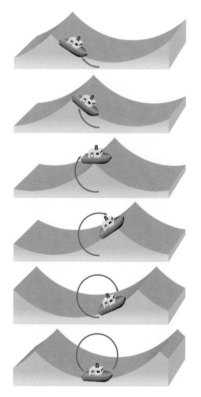

Figure 20.4 **The movement of a wave.**

Tarbuck/Lutgens, *Earth: An Introduction to Physical Geology, 8e*
© 2005 Pearson Prentice Hall, Inc.

NOTES:

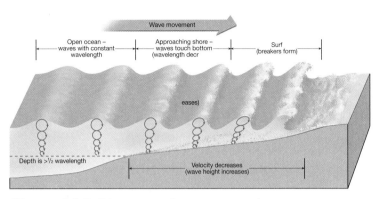

Figure 20.5 Changes that occur when a wave moves onto shore.

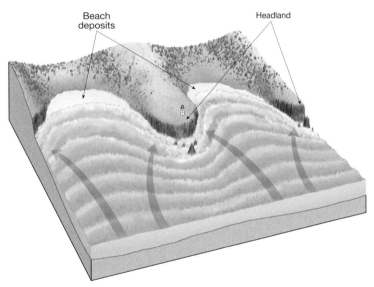

Figure 20.9 Wave refraction on headlands.

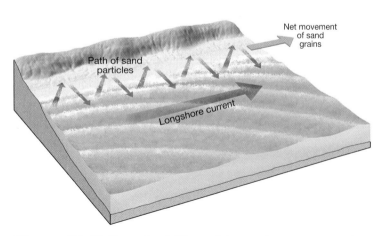

Figure 20.10 Beach drift and longshore currents.

© 2005 Pearson Prentice Hall, Inc.

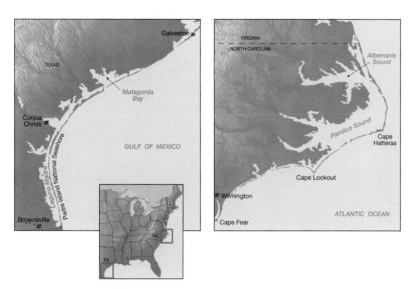

Figure 20.14 Barrier islands along the Gulf and Atlantic coasts.

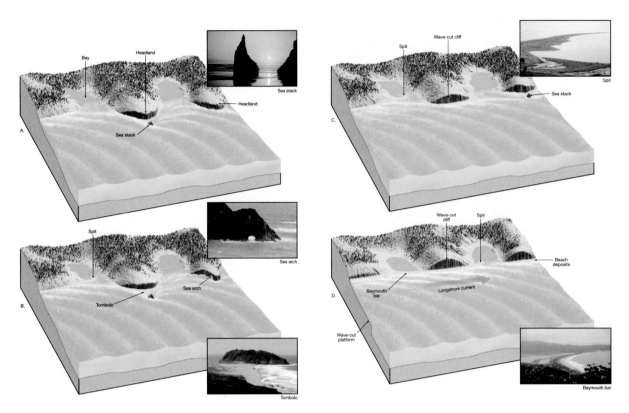

Figure 20.15 A,B,C,D *Changes along an initially irregular coastline.*

NOTES:

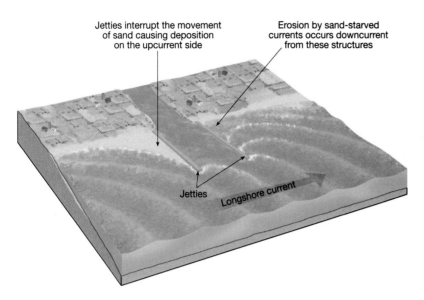

Figure 20.17 *Jetties.*

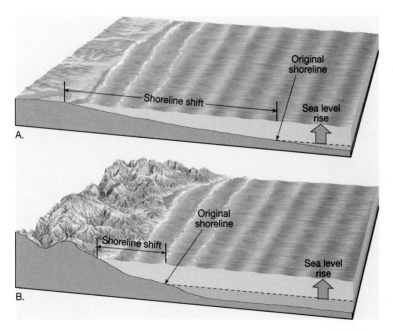

Figure 20.D A,B *Sea-level change and shore slope.*

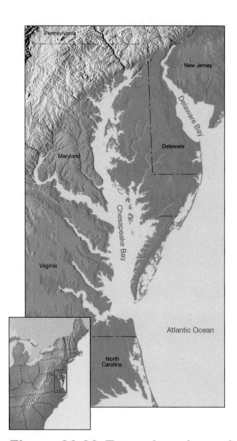

Figure 20.22 Estuaries along the East Coast of the United States

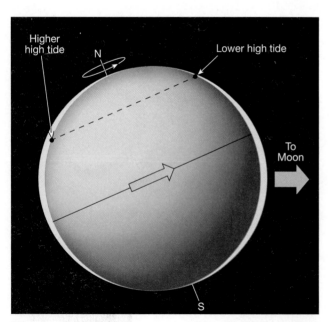

Figure 20.24 Tides on an Earth with uniform water depth.

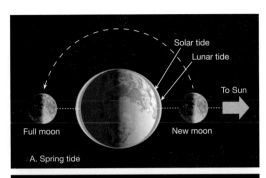

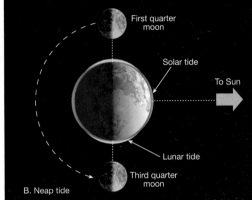

Figure 20.25 A,B Spring tides and neap tides.

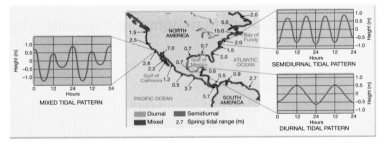

Figure 20.26 Tidal patterns.

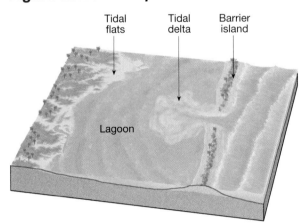

Figure 20.27 Tidal delta on the landward side of a barrier island.

Tarbuck/Lutgens, *Earth: An Introduction to Physical Geology, 8e*

© 2005 Pearson Prentice Hall, Inc.

NOTES:

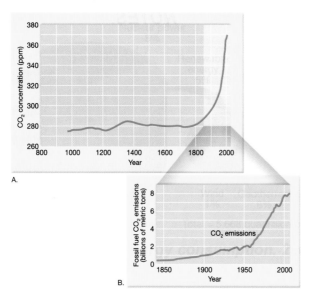

A.

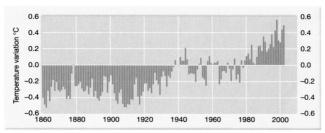

B.

Figure 21.10 A,B *Carbon dioxide concentrations over the past 1000 years.*

Figure 21.11 *Global temperature variations, 1860-2002.*

Figure 21.12 *North American oil sand deposits.*

Tarbuck/Lutgens, *Earth: An Introduction to Physical Geology, 8e*
© 2005 Pearson Prentice Hall, Inc.

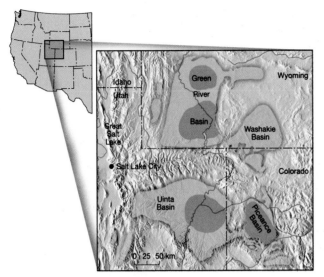

Figure 21.13 *Distribution of oil shale in the Green River Formation.*

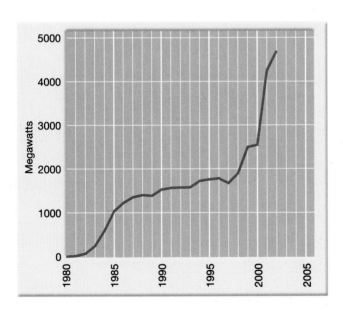

Figure 21.16 *U.S.-installed wind power capacity.*

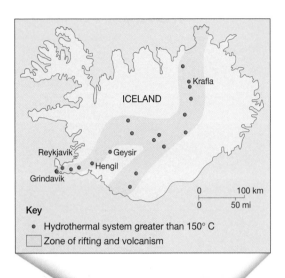

Figure 21.18 Iceland geothermal energy.

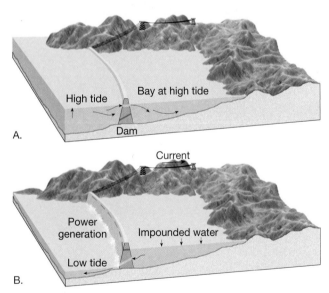

Figure 21.20 A,B The principle of the tidal dam.

© 2005 Pearson Prentice Hall, Inc.

NOTES:

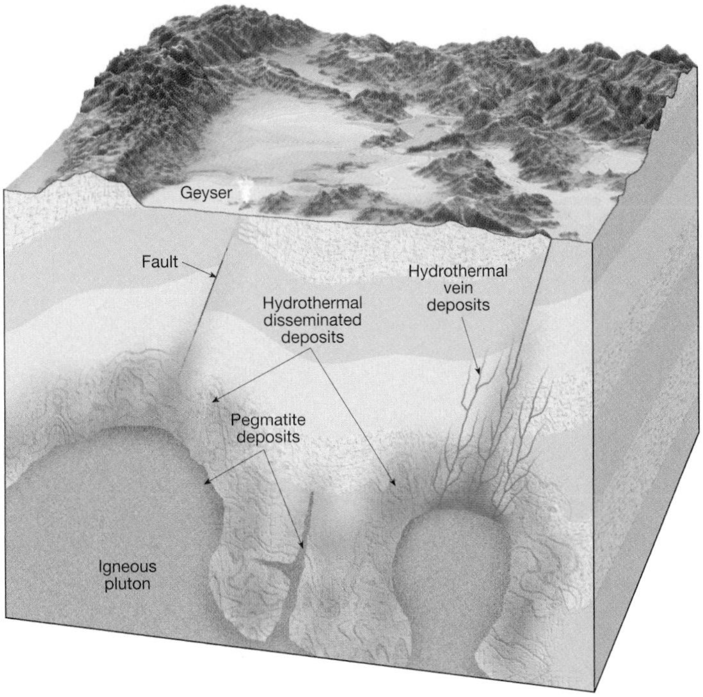

Figure 21.21 Pegmatite and hydrothermal deposits.

Figure 21.25 Seafloor sulfide deposits.

© 2005 Pearson Prentice Hall, Inc.

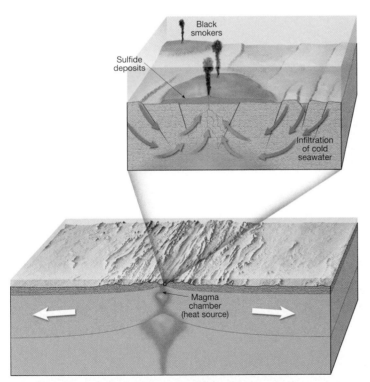

NOTES:

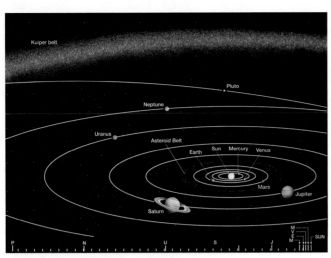

Figure 22.1 *Orbits of the planets.*

Figure 22.2 *The planets drawn to scale.*

Tarbuck/Lutgens, *Earth: An Introduction to Physical Geology, 8e*
© 2005 Pearson Prentice Hall, Inc.

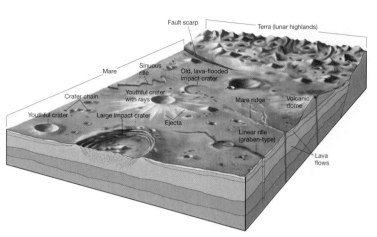

NOTES:

Figure 22.4 Major topographic features on the lunar surface.

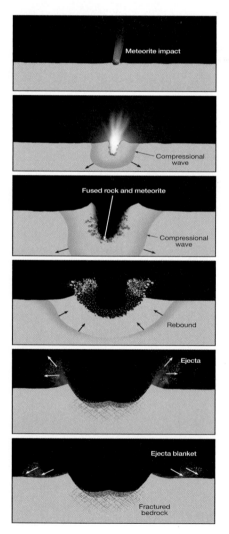

Figure 22.5 Formation of an impact crater.

© 2005 Pearson Prentice Hall, Inc.

NOTES:

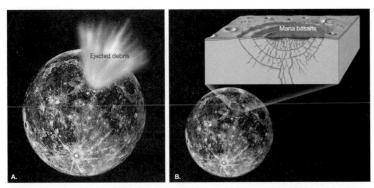

Figure 22.7 A,B Formation of lunar maria.

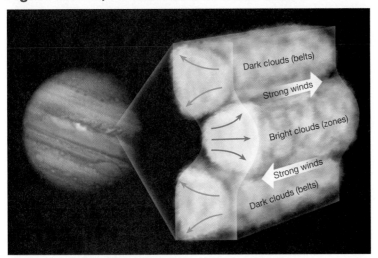

Figure 22.17 The structure of Jupiter's atmosphere.

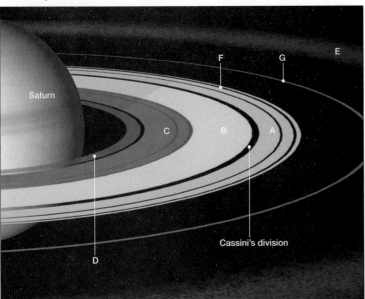

Figure 22.20 A view of Saturn's dramatic ring system.

Tarbuck/Lutgens, *Earth: An Introduction to Physical Geology, 8e*

© 2005 Pearson Prentice Hall, Inc.

NOTES:

Figure 22.B Pluto and its moon Charon.

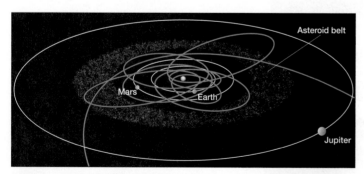

Figure 22.24 The orbits of most asteroids lie between Mars and Jupiter.

NOTES:

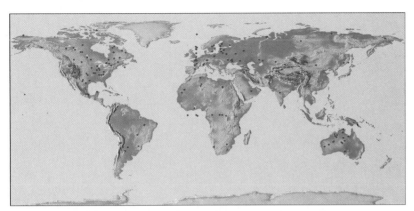

Figure 22.C World map of major impact structures.

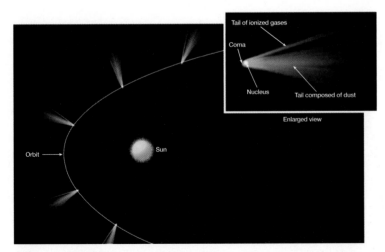

Figure 22.26 Orientation of a comet's tail as it orbits the Sun.

NOTES:

NOTES:

NOTES:

NOTES:

NOTES:

NOTES:

NOTES:

NOTES: